By JOHN BAECHTEL

HIGH PERFORMANCE CRATE MOTOR BUYER'S GUIDE

By JOHN BAECHTEL

EDITED BY
JOHN BAECHTEL

PRODUCTION BY
JOHN BAECHTEL

OVERSEAS DISTRIBUTION BY:

BROOKLANDS BOOKS LTD.
P.O. BOX 146, Cobham, Surrey, KT11 1LG, England
Telephone 01932 865051 • FAX 01932 868803

BROOKLANDS BOOKS LTD.
1/81 Darley Street, P.O. Box 199, Mona Vale, NSW 2103, Australia
Telephone 2 9997 8428 • FAX 2 9997 5799

ISBN 1-884089-13-5
PART No. 32

CARTECH, INC, 11481 KOST DAM RD., NORTH BRANCH, MN 55056

CONTENTS

ON THE COVER: Our cover engine is courtesy of Racing Head Service in Memphis, Tennessee. It is a 350 Chevy equipped with a polished 6-71 BDS supercharger, dual Holley carburetors, Competition Cams camshaft and valve train and MSD ignition. It comes with Manley stainless steel valves, RHS signature series valve covers, magnetic trigger distributor, Blaster -2 ignition coil, low compression blower pistons and heavy duty crank and rods. This engine comes fully assembled and crated with dyno testing available prior to shipping.

Left Photo courtesy of Mopar Performance

High Performance crate motors have become the powerplant of choice for street rodders, street machiners and other automotive enthusiasts seeking good performance at reasonable cost. Many engine shoppers don't want to bother with the myriad details of rounding up parts and assembling their own engines. There are numerous engine kits on the market, but crate motors make everything easy for the average enthusiast because they are fully assembled and guaranteed to make their advertised torque and horsepower ratings.

Basic engine assemblies have always been available from the major automakers, but only recently have they seen the advantages of offering competitively priced performance engines to meet the demands of brand loyal enthusiasts. Chevrolet, Ford and Chrysler have all developed HO performance engines for the enthusiast market and the high performance industry is literally bursting with reliable high performance crate motor suppliers.

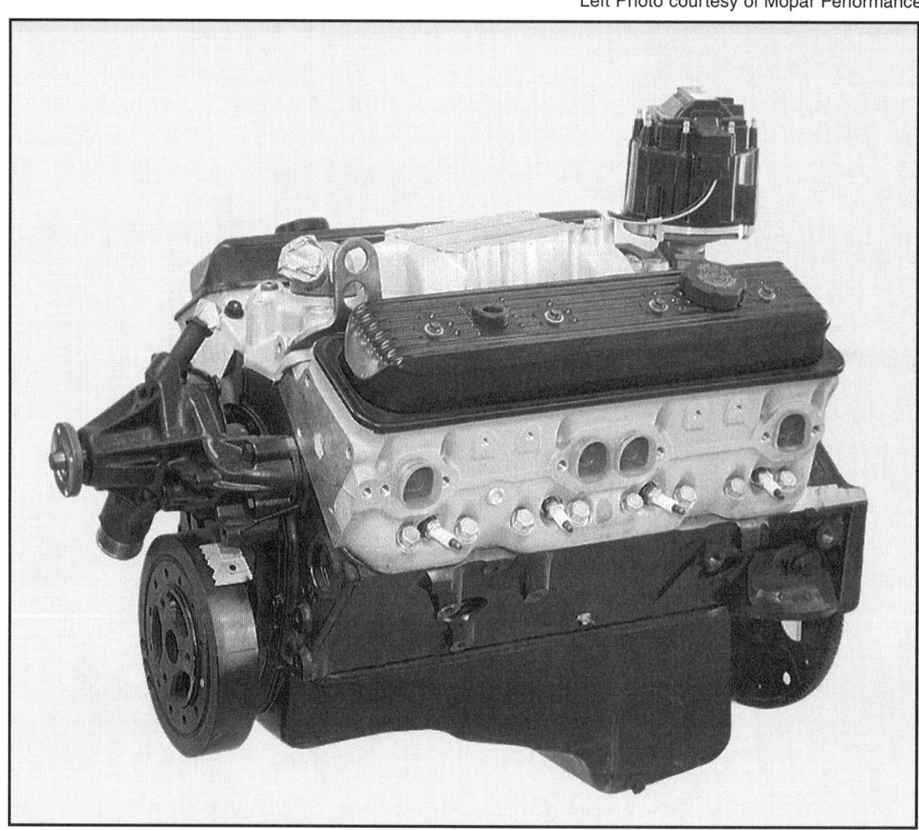

High performance crate motors like this 345 HP ZZ3 Chevy small block with aluminum heads have become the most popular choice for street rodders, street machiners and engine swappers in the nineties. Complete and semi-complete performance engines are available from the factory or from the high performance aftermarket.

There's no guesswork in a crate motor assembly because the combination has already been developed and tested. The customer doesn't have to worry about selecting the wrong camshaft or getting the compression ratio too high for pump gas. Most crate motor manufacturers are careful to validate their performance claims so you can be assured of getting a motor that performs as advertised. This doesn't mean that you can never get a lemon in a crate, but in most cases the crate motor you buy will deliver the power level and reliability you are seeking.

The High Performance Crate Motor Buyer's Guide is intended to give you an in-depth look at the most popular crate motors so you can decide for yourself. It will also show you how to make the best selection and categorize all the essential components you will have to supply in order to install the engine and get it running in the vehicle of your choice.

THE CRATE MOTOR CONCEPT

Crate motors come in various stages depending on the manufacturer. The best crate motors are complete from oil pan to valve covers, including the intake manifold. Most factory crate motors are production line assembled, which means they receive the same select fit components applied to all production auto engines. Factory crate motors typically offer one horsepower per cubic inch with factory engineered components that pass stringent production durability tests, and they carry a full factory warranty. They are also designed to meet current emission requirements when installed with all the appropriate accessories.

Factory crate motors like this 302 GT-40 Ford small block have become increasingly popular for both carbureted and fuel injected applications.

Most high performance industry crate motor assemblies offer one horsepower per cube and more. They come in various stages of completion with broad choices regarding induction systems. Some engines are complete assemblies while other are long block assemblies that have no intake manifold or distributor. Some factory engines are so complete they even come with factory headers, spark plugs, ignition wires, distributor, carburetor and air cleaner. This makes them an attractive package for engine swappers seeking the least painful way to put some grunt in their street rod or street machine.

The typical crate motor will be lacking a water pump and distributor, and not many come with a harmonic damper, flywheel or flex plate. Some are offered without intake manifolds, but most come with valve covers which you may or may not like. Additionally you'll need items such as motor mounts, spark plugs, ignition wires, headers, fuel lines, starter and other accessories to get your engine up and running in the vehicle. Depending on the vehicle or the application, you will probably already have some of these components from the previous engine. This can make your crate motor experience easier but not necessarily fool proof or trouble free.

Crate motors can take the worry and hassle out of building a hot street car because they eliminate selecting and chasing down parts, getting the machine work done and performing tedious assembly work. You can add your own choices for the induction, exhaust and ignition systems and get a personalized combination.

FACTORY vs AFTERMARKET CRATE MOTORS

If you choose a factory crate motor, you are getting an assembly that has been engineered and tested in the same manner as a production engine. This means that it meets all of the

Aftermarket crate motors generally offer more variety in terms of displacement, power ratings and high performance equipment content.

strict performance and durability levels that the factory deems necessary to ensure trouble free operation for the length of the warranty period and well beyond that with proper maintenance. Factory engines have to meet stiff requirements regarding power output, piston fit, ring and valve sealing, oil consumption and component longevity. They are assembled with all new parts and carry generous warranty terms. Thousands of these engines are on the road delivering strong, reliable performance for reasonable cost.

Crate Motor Applications

- **Street Rod**
- **Daily Driver**
- **Off-road Vehicle**
- **Budget Bracket Racer**
- **Show Car**
- **Shop Truck**
- **Street Machine**
- **Tow Vehicle**
- **Marine**
- **Low Rider**
- **Cruiser**
- **Kit Car**

Performance aftermarket crate motors also deliver good performance. While most factory engines are assembled with modern hypereutectic pistons, aftermarket engines are offered with cast pistons, hypereutectic pistons and forged pistons depending on the supplier. Some of these engines offer more than one horsepower per cubic inch and they are built with quality aftermarket performance parts. While factory quality control measures are more consistent, quality control in the aftermarket is at the whim of the manufacturer. It is possible to buy a motor with good pieces that have been poorly matched and sloppily assembled. The major suppliers represented in this book generally supply high quality performance engines. Occasional lemons have been known to slip through both at the factory and in the performance aftermarket, but the quality is generally high. If you are not sure about something, work out the terms and get it in writing, or shop elsewhere. As you will see the available choices are plentiful.

POWER RATINGS

Factory engines are rated to include standard parasitic losses from the alternator, water pump and so on. That's why these engines usually deliver about one horsepower per cubic inch of displacement or slightly less depending on the application. Performance aftermarket engines are usually tested on the dyno with little or no parasitic drag from accessories. The power numbers quoted are mostly correct and often higher than conservative factory ratings. These engines may produce more power, but they are often less fuel efficient than factory motors and they may not deliver the drivability some enthusiasts are seeking. It is critical that you honestly evaluate your engine requirements before investing in any crate motor application. The rule of thumb is: Factory motors are usually smooth, fuel efficient and reliable with good low end torque. Performance aftermarket engines can offer the same characteristics when properly tuned, but they generally have more camshaft timing and sacrifice some drivability. There is a middle ground and both types respond favorably to performance enhancements such as nitrous oxide injection and supercharging. You have to study your options carefully before making a choice. Giving you all the facts is the real purpose of the Crate Motor Performance Buyer's guide.

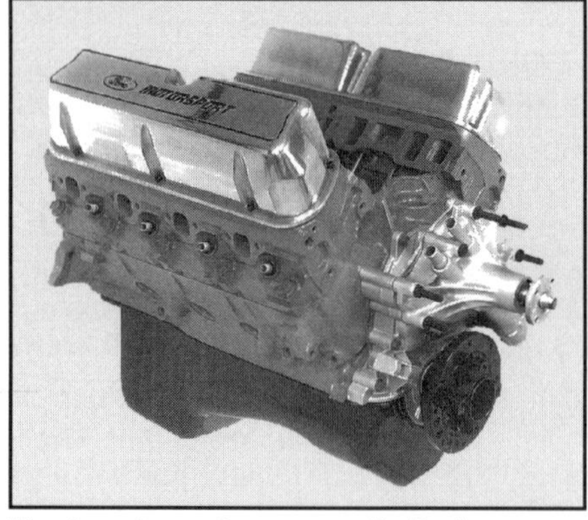

Most crate motors carry limited warranty terms and feature higher quality parts and workmanship than the average rebuilder engines available locally.

FUEL REQUIREMENTS

All factory performance crate motors are designed to deliver maximum performance from pump gasoline. Compression ratios range from 8.5:1 to 9.5:1 and the cylinder heads

CRATE MOTOR Performance Tips & Tricks

Advantages and Disadvantages

ADVANTAGES

- **Broad factory & aftermarket availability**
- **Limited warranties**
- **Guaranteed power levels**
- **Competitive pricing**

DISADVANTAGES

- **Few totally complete packages**
- **Most require additional components**
- **Limited warranties**
- **Not usually emissions legal**

State Emission Programs

The following are current 1995 state emission requirements. These are subject to change at any time, but you can use them as general guidelines for determining the smog checks that may apply to your vehicle. Some states do not have smog programs and many states only have them in selected high density areas. If you don't live near a large city you may not be subject to any smog programs. If your vehicle will pass the indicated tests and visual checks you should be home free. Be sure to keep a record of any engine parts you are using that carry Executive Order exemption numbers or EPA legal certification. Some inspectors may not be aware of the exempted parts and they may fail your vehicle unless you can prove that the parts are smog legal.

If you are installing a crate motor in an existing chassis the inspectors will first check the Vehicle Identification Number (VIN) to determine the vehicle's basic requirements. If the engine is a later engine they will probably require you to upgrade the emissions controls to match the engine. If it is an earlier engine it will usually have to match the requirements of the chassis. Most inspectors will try to get you passed or tell exactly what you need to do. It is easier to qualify for a waiver with an earlier model car, and cars that are subjected to basic tests have lower requirements and a lower dollar repair amount than those which are required to pass an IM240 test. The IM240 test requires you to spend up to $450 on repairs before being granted a waiver.

STATE	REQUIREMENTS	FREQUENCY	YEARS TESTED	MAX FEE	WAIVERS	TEST	ENFORCEMENT TYPE	VISUAL CHECKS
Alaska	Test & Repair	Annual	1975+	$50	$450	2	RD	A,C,E,H,I,P,W
Arizona	Test Only	Annual	1967+	$5.75	$300	1,3	RD	A,C,I,L
California	Test & Repair	Biennial	1966+	$37	$50	1,2	RD	A,B,C,E,I,V
Colorado	Test & Repair	Annual	1968+	$9	$200	1,2	RS	A,C,I
Connecticut	Test Only	Annual	1968+	$10	$40	1	S	on waiver only
Delaware	Test Only	Biennial	1968+	None	$75	1	RD	None
Dist. of Columbia	Test Only	Annual	1968+	$5	value of car	1	S	None
Florida	Hybrid	Annual	1975+	$10	$100	1	RD	C,G,I
Georgia	Test & Repair	Annual	-12 yrs.	$10	$50	1	RD	A,C,I
Illinois	Test Only	Mixed	1968+	None	Tune up	1,2	LS	on waiver only
Indiana	Test Only	Biennial	1976+	None	$75	1,2	RD	C,G,I
Kentucky	Test Only	Annual	1968+	$8.25	$15	1	SP	None
Louisiana	Visual Only	Annual	1980+	$10	None	None	S	A,C,G,I,L,V
Maine	None	—	—	—	—	—	—	—
Maryland	Test Only	Biennial	1977+	$8.50	$150	1	RS	C,I
Massachusetts	Test & Repair	Annual	-15 yrs.	$15	$100	1	S	C,G,I
Michigan	Test & Repair	Annual	-9 yrs.	$10	$82	1	RD	on waiver only
Minnesota	Test Only	Annual	1976+	$8	$75- $200	1	RD	C,G
Missouri	Test & Repair	Annual	1971+	$10.50	Tune-up	1	RD	A,C,V
Nevada	Test & Repair	Annual	1968+	$22.50	$100	2	RD	A,C,I
New Hampshire	Test & Repair	Annual	-15 yrs.	Market	$450	1	S	C
New Jersey	Hybrid	Annual	All years	None	None	1	S	C,I
New Mexico	Test & Repair	Biennial	1975+	Market	$75- $200	1,7	RD	A,C,I
New York	Test & Repair	Annual	All years	$17	Tune-up	1	S	A,B,C,E,F,I,V
North Carolina	Test & Repair	Annual	1968+	$19.40	$50	1	S	A,B,C,E,F,G.I.O.V
Ohio	Hybrid	Annual	1975+	$8	$100- $200	1,7	RD	A,C,E,G,I,V
Oregon	Test Only	Biennial	1975+	$10	None	2	RD	A,B,C,E,G,I,V,
Pennsylvania	Test & Repair	Annual	All years	$8.48	$25- $50	1	S	on waiver only
Rhode Island	Test & Repair	Annual	1967+	$15	None	1	S	C,I
Tennessee	Test Only	Annual	-12 yrs.	$6	None	1	RD	G,I
Texas	Test & Repair	Annual	1968+	Mixed	$200- $250	1,7	S	C,I
Utah	Test & Repair	Annual	1968+	$14	$100- $200	2	RD	A,C,E,G,I
Vermont	None	—	—	—	—	—	—	—
Virginia	Test & Repair	Biennial	-20 yrs.	$13.50	$200	1	RD	A,B,C,I,S,V
Washington	Test Only	Biennial	1968+	$12	$100- $150	1,3	RD	All Primary ECD
Wisconsin	Test Only	Annual	1968+	None	$75- $200	1	RD	A,B,C,E,F,G,I

Legend

Tests Performed:
1 Idle
2 2-speed
3. Loaded/idle
4. IM240
5. Canister purge
6. Pressure
7. Lead

Enforcement:
RD Registration denial
R Registration
S Windshield sticker
LS License suspension
RS Registration suspension
SP Subpoena

Visual Checks
A. Air pump
B. Evaporative canister
C. Catalyst
E. EGR
F. Air cleaner
G Gas cap
H. Hoses
I. Inlet restrictor
J. Vacuum lines
O. Oxygen sensor
P. Panel lights
S. visible smoke
V. PVC
W. Wires

incorporate factory hardened seats for compatibility with unleaded fuel. These are truly plug and play engines that should deliver strong consistent power without detonation or poor drivability as long as you use high quality 92 octane pump gas and maintain proper tune up specs.

Most aftermarket engines are the same, but some are still built with rebuilt heads that are not compatible with unleaded fuel. Some also come with higher compression ratios which will knock and incur engine damage on unleaded pump gas. When you order a crate motor make sure you ask about fuel compatibility with regard to detonation and valve seat recession—the two most likely things that can turn a new crate motor into a basket motor.

Buyers frequently have a choice between cast iron and aluminum cylinder heads. Iron is more detonation prone and aluminum is easier to port and repair. Match your cylinder head selection to give your application the most advantages.

CRATE MOTOR WARRANTIES

Most crate motors are covered by some type of limited warranty, but engine builder's warranty policies differ so you need to get the specifics from your chosen supplier at the time of purchase. Major engine builders such as Summit Racing offer warranties that generally cover the entire engine for the first 90 days, but some suppliers offer no warranty coverage because they consider performance engines the same as racing engines. Always get the specifics of the warranty before making your decision.

All engine builders will want to examine a failed or damaged engine prior to complying with any warranty they provide. The customer is usually required to pay for the shipping charges, but the engine builder will cover the shipping for a replacement engine should one be necessary.

Of course the customer is expected to exercise proper care in the installation and initial break-in of the new engine. If you start the engine without oil, or seriously overheat it, the clues will be evident and the engine builder may refuse to honor the warranty. Most engine suppliers understand that you are buying a high performance engine because you want to drive it hard. If they have done something wrong, they will usually fix it or replace it, but you have to bear the responsibility if you ruin the engine as a result of careless mistakes.

CRATE MOTORS AND EMISSION LAWS

Both forged and steel crankshafts are available. If you plan to lean on the motor with RPM, boost pressure or nitrous, always choose a forged crank.

Very few high performance crate motors are emissions legal, yet they are routinely installed in all types of vehicles which have various emission requirements. If the engine is installed with the appropriate emissions control equipment hooked up and operating, it may meet emission requirements in most states. In some states (California for example) you can take the vehicle to a referee station and have it inspected and certified if it passes their requirements.

The key to maintaining emissions legal status is to avoid downgrading the engine and emission controls systems for the year of the chassis in which the engine will be installed. For engine swaps, most states require that the combination meet the standards for whichever is newer—either the chassis or the engine. The state can require your car to meet either standard depending on which standard ensures optimum compatibility of the electronic controls if applicable.

The EPA has its own guidelines which apply to engine swaps and replacement engine changes. Specific requirements are also spelled out in the IM-240 regulations which states must use to implement inspection and maintenance programs. Contrary to popular misconception it is desirable to inform an experienced inspector about your engine combination so he can help you determine the best course of action to ensure its legality.

You cannot make modifications that increase emissions, but you can use any components that have been certified or exempted in California, comply with EPA regulations or are accepted as actual replacement parts. To determine this you have to ask if the parts have a California Air Resources Board (CARB) Executive Order (E.O.) number or that they meet the requirements of the EPA's anti-tampering document Memorandum No. 1A.

Many crate motors will have some legal parts and some illegal parts. You have to quiz the engine supplier carefully to determine the true legality of the engine when installed in your vehicle. Sometimes minor changes are all that is required to bring the engine

The first step in creating a powerful and reliable crate motor is hot tank cleaning, inspection and pressure testing to determine the suitability of the cylinder block for use in a high performance application.

Torque plate honing has nearly become an industry standard for high performance crate motors. This procedure ensures dimensionally correct cylinder bores under actual operating conditions.

into compliance, but in other cases the engine may be completely illegal. It is critical that all required emission control devices are installed and operating properly. This means catalytic converters, EGR valves, Oxygen sensors, smog pumps, evaporative emissions canisters and whatever else applies.

For most crate motors you may not be able to determine the exact year of the engine so the year of the chassis may determine your requirements, but not always. When in doubt try seeking out a qualified state inspector and ask for a specific ruling on your proposed combination. You may have to alter it slightly, but in many cases they will give you the guidance you need to

CRATE MOTOR Performance Tips & Tricks

Piston Materials and Applications

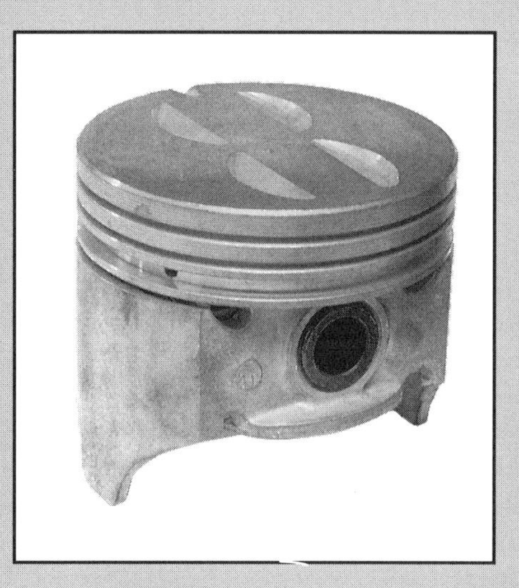

Cast Pistons have been the standard automotive piston for decades. They are inexpensive and easy to produce. They have a thermally stable crystalline grain structure and often incorporate cast-in steel expansion struts that allow them to fit tightly in the bore for optimum stability and ring seal. Under normal use they will stand up well to tens of thousands of miles of use. However they have limited speed, thermal and detonation resistance. They should only be used in in limited performance engines where engine speed is limited and detonation is strictly avoided.

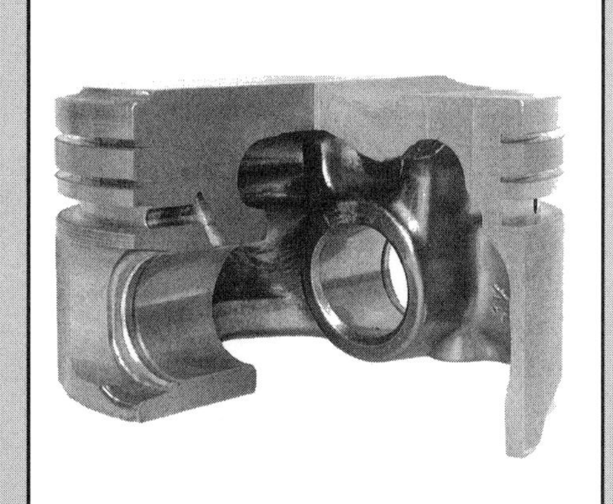

Hypereutectic® pistons are also castings, but they have almost 2-1/2 times the silicon of a standard cast piston for increased hardness and greater resistance to higher temperatures and cylinder pressures. They are dimensionally stable and require very little skirt clearance. In some instances they can be operated with less clearance than standard cast pistons. This feature keeps the piston and ring package well stabilized in the bore and improves sealing and blowby control. While Hypereutectic pistons are well suited to street performance applications, they still lack the detonation and temperature resistance of forged pistons. They should not be used with any more than light loads of nitrous-oxide injection, or with high pressure turbocharging or supercharging systems. Add on turbocharging or supercharging systems should be limited to 6 to 8 pounds of boost to avoid piston damage. These pistons are tougher than cast pistons, but they don't approach the strength afforded by true forged pistons.

Forged pistons are manufactured with a forging die from a solid slug of heat-treated aluminum alloy. They possess the dense grain structure and metallurgical properties to stand up to severe use, including a degree of detonation resistance. However forged pistons have less dimensional stability and require greater skirt clearance for reliable operation. Forged pistons remain the top choice when strength and durability are required for racing, turbocharged, supercharged or nitrous-oxide injected applications.

ensure a street legal crate motor application.

HOW TO DETERMINE YOUR NEEDS

When you consider a crate motor for your vehicle it important to accurately define your needs. What do you really want? How will you use the vehicle? How much drivability and fuel economy are you willing to sacrifice in exchange for brute horsepower? You have to decide whether you are looking for low speed torque or high speed power. Do you want a broad, flat torque curve or a narrow range with peak power at high rpm? How often will you drive the vehicle and what type of driving do you anticipate. Don't mislead yourself. You will be much happier with your decision if you restrain yourself from selecting a combination that is too wild.

Bore spacing is adjusted during the boring process to bring all cylinder bores into proper alignment and perpendicular to the crank centerline.

Modern automated honing equipment ensures that every cylinder is finished exactly the same so you can enjoy optimum ring seal and maximum power.

Cylinder blocks are aligned honed to true the crank centerline within the main bearing housing bores. A choice of either 2-bolt or 4-bolt mains is usually made available to the customer.

Decks are surfaced 90 degrees from each other and perpendicular to the crankshaft centerline. Each deck is machined equal distant from the crankshaft and finished with the optimum surface to ensure good gasket sealing.

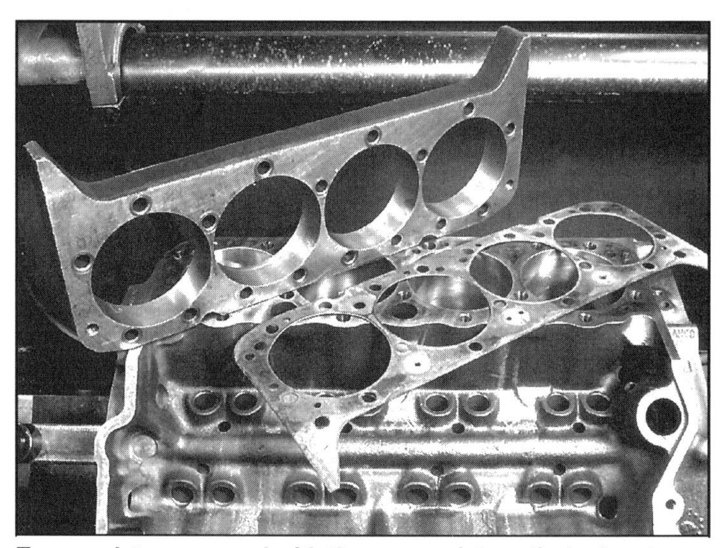

Torque plates are used with the appropriate cylinder head gasket to duplicate cylinder wall distortion caused by the bolt torque applied to the cylinder head bolts.

Some of the most important considerations include engine size, stroke length, intake configuration, carbureted or fuel injected, exhaust system, type of transmission and rear end gearing. Is fuel economy important? What fuel quality can you provide for the engine? What type climate will you drive in? Do you need to consider cold weather drivability? Will the engine require additional cooling capacity to handle air conditioning and excess engine heat in a crowded engine compartment? All of these and other factors will affect your choice. It is a good idea to write down every factor you can think of that will affect your choice and keep the list handy when you start consider the broad range of crate motor applications that are available.

HOW TO ORDER A CRATE MOTOR

Factory crate motors can be ordered through the parts department of any established dealership anywhere in the country. Most dealers use a priority system that attempts to locate the part or the engine at the dealership first, then it checks the local distribution center and then moves on to a regional distribution center if necessary. Stock, or "PAD" orders are easiest and most economical, but the price of

Some crate motors builders have their own line of connecting rods while others use either new rods or reconditioned rods. Either approach is acceptable, but the application should dictate the type of rod you select. Used rods are Magnafluxed to check for cracks and the beams are polished to eliminate stress points. The most important points are the quality of the rod bolt and the final finishing of the big end. Always insist on top quality rod bolts.

your engine should not change if it is not available locally. The only difference should be shipping priorities which may delay delivery.

Aftermarket engines are just as easy to obtain. you can order by phone and have the engine delivered to your door via truck within a few days. You are generally responsible for the shipping charges and the purchase may be handled C.O.D. or by credit card.

When ordering a performance aftermarket crate engine you have more choices and the salesman may be able to offer you more options based on your stated requirements. If you can answer most of the questions on the accompanying checklist, the salesman can probably get you an engine that is closely matched to your requirements.

It is very important to keep a careful record of all your transactions in case there is a problem later. Establish a file or an envelope to hold paperwork such as copies of the order form, shipping papers, delivery receipt, C.O.D. forms, checks and any other paperwork associated with the engine. This file should also contain a complete description of the engine, a copy of the build sheet and specifications for parts like the camshaft. These can be helpful for tuning purposes. It is a good idea to take some photographs of the engine when it is delivered. This will establish its condition at delivery and serve as a visual record if it is ever needed for insurance purposes.

You may even wish to make a logbook on the engine so you can track performance and expenses. The book should contain entries for routine maintenance such as oil and filter changes, part numbers for filters and types of lubricants, valve lash settings, timing settings, a bolt torque guide and room for notations about anything that may occur during the life of the engine. While this may seem a little much, you may be surprised how handy it becomes when you have to return or repair the engine or order replacements for a damaged component. The more you know about your crate motor, the better you'll be able to maintain and enjoy it for a long time. A little forethought here will save you time and money later.

HOW CRATE MOTORS ARE BUILT

Crate motors vary in quality depending on the manufacturer. The distinguishing factor that separates crate motors from run-of-the-mill rebuilder engines is high performance intent. Crate motors are intended for high performance applications. They are built with a healthy mix of performance parts designed to produce more power with good reliability.

Most of the major suppliers are well equipped to manufacture engines and they apply many of the most common high performance modifications when preparing crate motors for final assembly. This includes a thorough cleaning and inspection of each cylinder block. A large percentage of crate motors are actually built with a mix of new and rebuilt parts. Good ones have new cranks and rods in addition to new pistons, and extra effort is made to ensure accurate machine work. Where rebuilder engines are typically bored and honed to a relatively broad tolerance range, crate motors often received a considerable amount of block preparation and torque plate honing to ensure durability and optimum power output.

Some crate motors start life with a new cylinder block, but most are built with seasoned blocks that have already taken a set and will produce better power once they have been accurately machined and assembled with quality parts. These blocks are typically degreased and pressure tested to ensure quality. Some builders dip the blocks in chemical baths to remove rust and other deposits. All old core plugs and gallery plugs are removed to ensure thorough cleaning and all of the bolt holes are retapped to ensure clean threads.

You usually have a choice between

Performance aftermarket crate motors are assembled with care and quality parts. This engine builder is checking crankshaft end play on a Chevy small block that will soon put 450 HP under some happy car owners hood.

All of the major auto manufacturers offer high quality, powerful crate motors designed for replacement service and high performance applications.

two or four bolt main bearings and some shops will also install race style nodular iron or steel main caps with splayed bolts at your request. The main bearing bores are align honed and the decks surfaced to make the decks parallel and equal distant from the main bearing centerline. Deck surfaces are machined 90 degrees from each other and finished for proper gasket seal. Bore spacing is corrected and some of the more thorough shops will square the back of the block with the crankshaft centerline. Boring and torque plate honing is performed carefully to ensure optimum bore finish and piston skirt clearance. New core plugs and threaded gallery plugs are installed along with new cam bearings to get the block ready for final assembly.

Crate motor shoppers have different needs and the manufacturers try to meet these needs by incorporating all of the extra machining and critical dimension checking necessary to guarantee a powerful and reliable engine. Most of these steps incur additional expense, but the engine builder can still wrap it into an attractive price because he is mass producing many engines with the same basic specifications.

There are of course, varying levels of crate motor preparation and most engine builders will gladly specify the steps they perform on their engines. Some engine shops use high performance engine components, but they only employ standard engine rebuilder boring and honing techniques. This method is adequate to ensure a good running engine, but may not derive every last bit of performance available from the selected combination of parts. Most of the top crate motor engine suppliers incorporate all the standard high performance engine building tricks. Many smaller shops do the same, but it is a good idea to get the specifics so you know how your money is really being spent.

Keep in mind that it isn't necessarily a bad thing if the engine builder doesn't use torque plates and all the high performance block truing techniques. The engine will still perform well, but you will never know how much more power is locked inside the combination. The performance difference between a blueprinted and non-blueprinted engine can be considerable in some cases and very slight in others. Sometimes the additional blueprinting is available, but you have to ask for it and it costs extra. In almost every case it is worth the money, so don't let a few extra hundred dollars keep you from getting all the power and performance possible from a good crate motor.

Choosing a crate motor to suit you

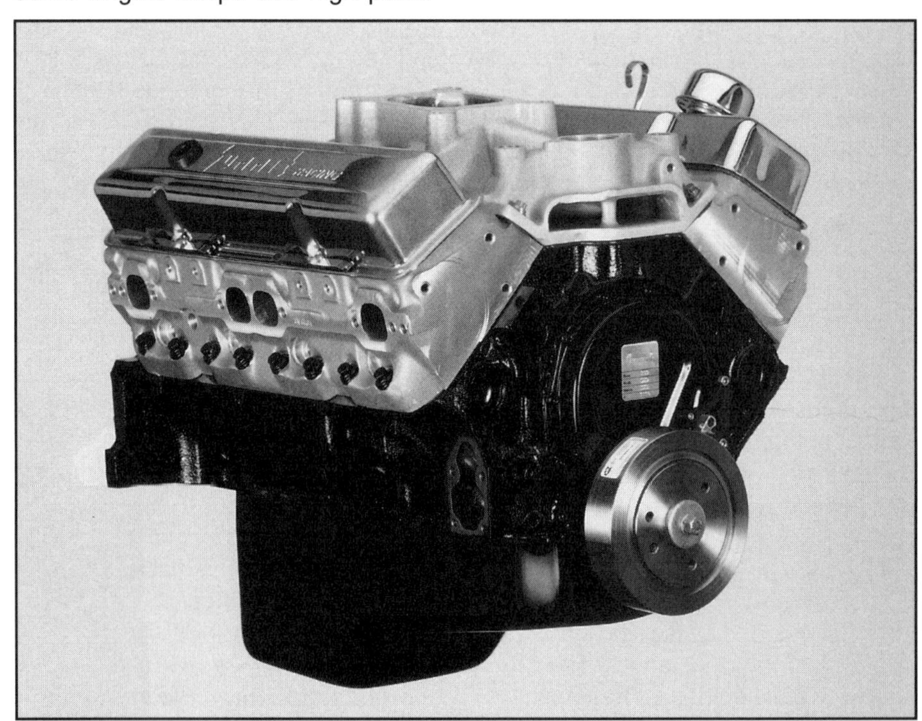

Aftermarket crate motors are usually very complete and they are built from quality components with extra careful machine work. Lemons are pretty rare within the aftermarket crate motor industry.

particular application isn't difficult, but it does require consideration of a lot of important points. The Crate Motor Buyer's Guide highlights the important elements of all the most popular domestic crate motors available from both the factory and the performance aftermarket. There are horsepower and component levels to suit every conceivable application. With so many choices it should be easy for you to select an engine combination that closely fits your needs and your carefully planned budget.

Between the major automakers and the performance aftermarket there are few applications that aren't covered by the crate motor movement. Detroit automakers are particularly driven to provide crate motor combinations that incorporate the latest design features of production engines because most of their crate motors are actually built on the production line.

Wherever possible the Crate Motor Buyer's guide will show you the most current setup from each manufacturer, and the previous setup that may still be sitting in some dealer parts departments and/or regional distribution centers. In many cases the changes are minor and incremental so the absolute most current version may not offer any power advantage, but rather a convenience for the manufacturer that allows them to continue offering high performance engines at a reasonable cost.

The aftermarket crate motor industry is keeping pace with modern OEM engine applications. In addition to carbureted crate motors, they can supply all types of crate motors equipped with the latest electronic fuel injection, plus superchargers, turbos and just about anything you might require.

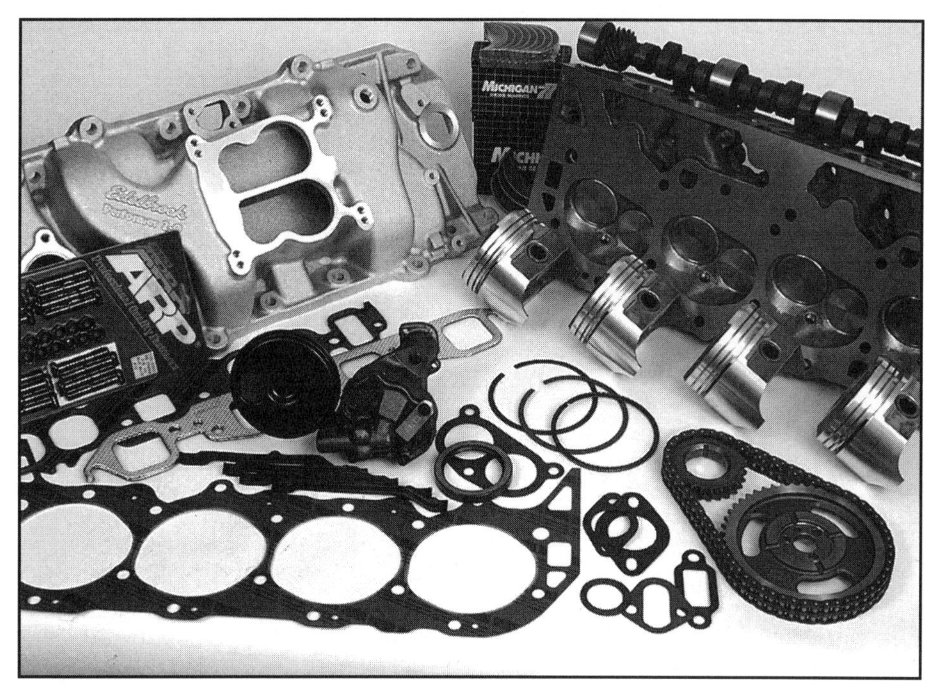

Aftermarket engine kits are a popular alternative to completely assembled crate motors. They are attractive for hardcore enthusiasts who wish to enjoy assembling their own engine.

CRATE MOTOR Performance Tips & Tricks

Engine Selection Checklist

- Type of vehicle ❑
- How is the vehicle used ❑
- How frequently is the vehicle driven ❑
- Drivability and comfort level ❑
- Cold weather drivability ❑
- Desired fuel economy ❑
- Available fuel quality ❑
- Desired displacement ❑
- Desired compression ratio ❑
- Desired power level ❑
- Desired torque peak ❑
- Desired power peak ❑
- Desired idle quality ❑
- Engine speed range (RPM) ❑
- Type of camshaft ❑
- Aluminum heads or iron heads ❑
- Valve seats compatible with unleaded fuel ❑
- Steel crank or cast crank ❑
- Carbureted or electronic fuel Injection ❑
- Standard block or heavy duty ❑
- Type of piston material ❑
- State of engine completion ❑
- Type of transmission ❑
- Rear gear ratio ❑

Illustration courtesy of Chevrolet

350 HO, 345 HP ZZ3 Engine Assembly

High Output 345hp, 350cid (5.7L—HO 350), PN 10185072, ID Code ZZ3. This 350 high-performance GM engine assembly produces 345hp and is designed for pre-'68 vehicles or off-road use (see photo in 350HO Camaro Conversion Package on page 16). The ZZ3 is a bolt-in replacement for any 265 through 400 smallblock engines and incorporates the latest high-performance components, including lightweight aluminum cylinder heads, low-friction hydraulic roller tappets, and hypereutectic (high silicon) aluminum pistons. All ZZ3 engine assemblies are test fired and final balanced at the factory to ensure top quality. They feature a new low-profile, dual-plane intake manifold (PN 10185063) for increased hood clearance without power loss. The manifold has a dual-pattern flange that accommodates both standard-flange Holley and spreadbore style Quadrajet carburetors, and it incorporates all accessory brackets, EGR (exhaust gas recirculation) port, and integral hot air choke (Block-off plates are installed on the EGR and choke stove on HO 350s).

Special features of the HO engine include lightweight valve spring retainers that have half the mass of the previous design, valve stem seals on the exhaust valves for enhanced oil control, and the aluminum heads are built with radiused valve seat inserts. This HO engine is equipped with stamped steel rocker covers that have PCV valve and oil filler holes on both covers. The engine is equipped with a torsional damper, a cast iron water pump (standard rotation, long style), a 12-3/4-inch, 153-tooth automatic transmission flexplate, and a dipstick. Spark plug wires, pulleys, starter

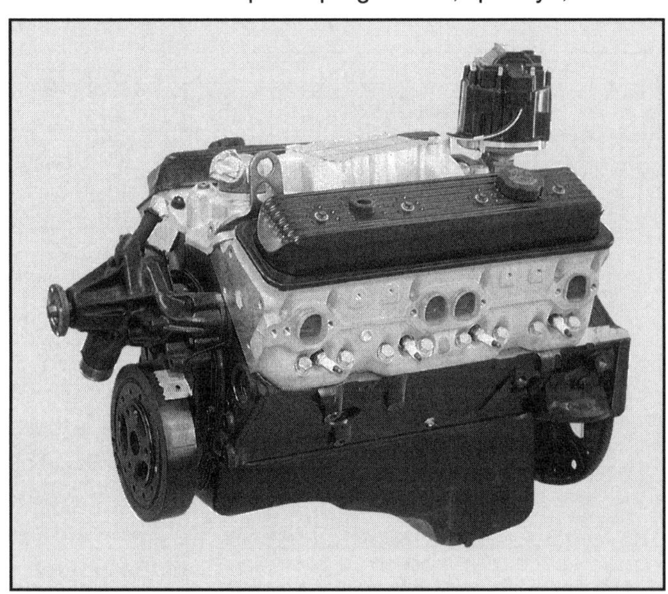

The ZZ3 HO small block crate engine is rated at 345 horsepower. It features a dual pattern hydraulic camshaft that delivers a broad, smooth torque range and good top end power for most street and highway applications.

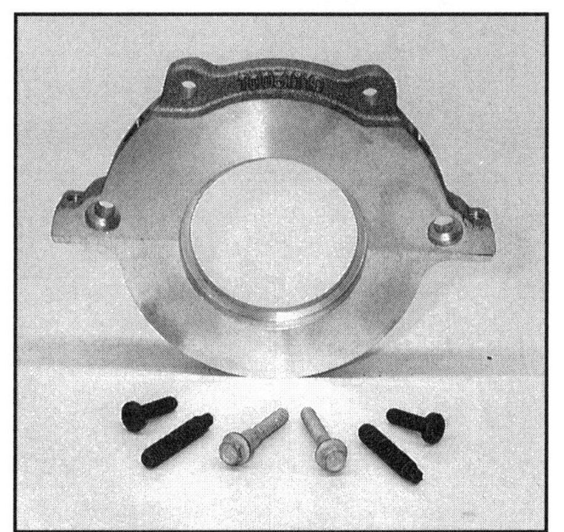

Since 1986 all production small blocks and crate motors have used a 1-piece rear main seal that mounts in an aluminum "adapter" (left photo). The adapter bolts into a machined recess at the back of the block (middle photo). Late crankshafts were redesigned for the 1-piece seal and incorporate a large round flange (arrow), instead of the earlier, large bolt pattern flange. Because of these changes, late cranks are not compatible with earlier blocks. However, early cranks can be used in late-model blocks when a GM crankshaft seal adapter PN 10051118 (right photo) and the early 2-piece seal is attached to the late block.

motor, oil filter, oil filter adapter, fuel pump, exhaust manifolds, and accessories are not included.

Two previous version of this engine, either ZZZ or ZZ2 may still be found in some dealerships. The following specifications are particular to these earlier engines.

The 1988 ZZZ crate motor PN10134338 is the original version of the 345 HP small block. It was built with cast 9.8:1 hypereutectic pistons. Unlike most production engines the pistons in this engine had no wrist pin offset. The engine featured a single pattern hydraulic camshaft with 235 degrees duration at .050-inch lift and .480-inch valve lift. A dual plane high rise intake manifold was standard and the engine was supplied with water pump, balancer, flexplate and HEI distributor.

Shortly thereafter a revised ZZ2 engine combination PN10185025 was introduced with off-set wrist pins and a few minor revisions to accomodate late model accessory drives.

All three of these engines are rated at 345 horsepower. The current ZZ3 engine exhibits the best drivability for most applications. ZZZ and ZZ2 engines had weaker low and mid-range torque, but they still offered plenty of power at higher engine speeds. They can be used with either manual transmissions or automatic transmissions with the appropriate flywheel or flexplate. Keep in mind that all flywheels and flexplates are neutral balanced with the exception of those used on 400 engines and late model 1986 and later engines with 1-piece rear seals. The correct late model flywheels and flexplates have a smaller 3.00-inch diameter crank flange bolt circle for use with the 1-piece real seal crankshafts. Both the HO and the SP 350 crate motors come equipped with the correct flexplate.

Flexplates have three different torque-converter bolt-pattern sizes. To determine the bolt pattern size on a torque converter, measure the distance from the center of a bolt hole to the center of the converter, then multiply this number by two. Refer to the accompanying chart on page 16 for application information.

Both flywheels and flexplates are available in large and small diameters. The larger, 168-tooth version is 14 inches in diameter, while the smaller, 153-tooth version measures 12-3/4 inches. To obtain proper starter engagement with 14-inch flywheels use the Chevy starter motor PN 1108400, and use starter motor PN 1108789 with the smaller 12-3/4-inch flywheels and flexplates.

And even when the right flywheel or flexplate is used, it can cause a lot of damage if the attaching hardware comes loose. Always use high quality fasteners (Grade-8 or better), and

345 HP HO 350 ID Code ZZ3

SPEC SHEET:

GENERAL:

Part Number	**10185072**
Displacement	350-cid, 5.7L
Bore & Stroke	4.00 x 3.48
Compression	9.8:1
Rated Power	345 @ 5400 rpm
Fuel	92 octane, unleaded premium

CAMSHAFT:

Part Number	**10185071**
Type	Hydraulic Roller
Valve Lift	.474/.510-inch
.050 Duration	208°/221° @.050
Int/exh lobe sep.	112°

CYLINDER HEAD:

Part Number	**10185086**
Material	Aluminum
Chamber Vol.	58cc
Valve Material	Steel
Int. Valve Dia.	1.94
Exh. Valve Dia.	1.50
Rocker Ratio	1.5
Valve Springs	PN 10134358
Intake Manifold	PN 10185063

SHORT BLOCK:

Block	4-bolt, PN 10105123
Crankshaft	1053 forged steel, 1- piece rear seal, PN 14096036
Rods	1038 forge steel, pink color code, PN 14096846
Pistons	Cast hypereutectic, PN 10181389

IGNITION:

Distributor	HEI, PN11033436
Spark PLugs	AC FR5LS
Plug Gaps	.035-inch
Timing	10° BTDC, 800 rpm

CRATE MOTOR Performance Tips & Tricks

HO 350 Camaro Conversion Package

PN 10185077

Chevrolet offers an 350 HO engine conversion package for 1982-87 Camaros originally equipped with LG4 or L69 engines and 700R4 automatic transmissions as **PN 10185077**. To identify your engine and determine the exact pieces needed to convert your vehicle, check the eighth character of the Vehicle Identification Number (VIN) as shown in the following sample VIN code:

1G1FP3IHOLL10001

If the eighth character is an "H," this indicates the vehicle is equipped with a 1982-87 LG4 305 cubic inch engine. If the eighth character is a "G", it indicates a 1983-86 L69 305 cubic inch engine.

When installed with the specified equipment (check with your Chevrolet dealer for a list of the required parts), this package meets EPA 49 state emissions standards under EPA Memorandum 1A. The conversion is also legal in California under California Air Resources Board Exemption Order #D-278.

In addition to the HO engine, the conversion package includes a dual-snorkel air cleaner with "ram air" ducts directing cool air to the carburetor. A high volume in-tank electric fuel pump is included along with vent lines, a sending unit, wiring harness and adjustable external fuel regulator. A low restriction exhaust system with dual-catalytic converters, high-flow exhaust manifolds with 2-1/4-inch outlets, a single 3-inch diameter tailpipe and transverse muffler are also part of the package.

To ensure proper calibration, the kit includes an engine control module (ECM), specially calibrated PROM chip, a coolant fan switch, electric spark control (ESC) module and Quadrajet secondary metering rods. The transmission gets a pressure regulator spring, boost valve, and servo assembly to improve shift firmness, and rear control arms with higher-durometer (50K) bushings are specified to reduce wheel hop.

During development of this package, a prototype HO engine equipped with a computer controlled (feedback) Quadrajet carburetor and full emissions controls delivered 308 hp at 5000 rpm and 365 lb-ft of torque at 3500 rpm. The Chevrolet Race Shop's 1987 Camaro testbed ran a 13.83 ET at 98mph in the quarter mile with the same basic package. Accomplished car builders may also want to select elements of this package to spec out powerful and fuel-efficient electronically controlled conversions for earlier pre-emission cars. Keep in mind that electronic fuel injection requires a minimum of 15 inches of vacuum for proper MAP sensor operation; hence you may wish to select an alternate camshaft profile that will not compromise engine idle vacuum.

The centerpoint of the HO350 Camaro Conversion package is the ZZ3 engine (see description on page 15). When installed with the specified components developed for this kit, this potent factory powerplant significantly improves vehicle performance while meeting EPA and CARB requirements for 50 states highway operation.

apply locking compound to the threads for an extra margin of safety. Torque all bolts to factory recommended specs (usually 60 foot pounds).

Factory crate engines usually come with the correct balancer and a flexplate for an automatic transmission. In the rare instance where you may wish to change these components, the accompanying charts will help ensure that you replace them with appropriate parts that are properly balanced and fit the engine correctly. You cannot ignore this requirement. If you use the wrong flexplate or balancer on an engine, serious damage may result from the imbalance condition.

Here is the 300hp, 350 SP engine as it looks when you open the crate. Just add your choice of intake manifold, carburetor and ignition, accessories and exhaust manifolds or headers, and fire it up! It comes dressed with chrome timing and rocker covers.

300 HP SP 350 ID Code SP

SPEC SHEET:

GENERAL:

Part Number	**12355345**
Displacement	350-cid, 5.7L
Bore & Stroke	4.00 x 3.48
Compression	9.1:1
Rated Power	300 @ 5000 rpm
Fuel	92 octane, unleaded premium

CAMSHAFT:

Part Number	**24502476**
Type	Flat Tappet Hyd.
Valve Lift	.435/.460-inch
.050 Duration	212°/222° @.050
Int/exh lobe sep.	N/A

CYLINDER HEAD:

Part Number	**12356026**
Material	Iron
Chamber Vol.	64cc
Valve Material	Steel
Int. Valve Dia.	1.94
Exh. Valve Dia.	1.50
Rocker Ratio	1.5
Valve Springs	PN 3911068
Intake Manifold	PN 14096011 or 10185063

SHORT BLOCK:

Block 10105123	4-bolt, PN
Crankshaft	Nodular iron, 1- piece rear seal, PN 14088527
Rods	1038 forge steel, PN 14031310
Pistons	Cast hypereutectic, PN 12350264

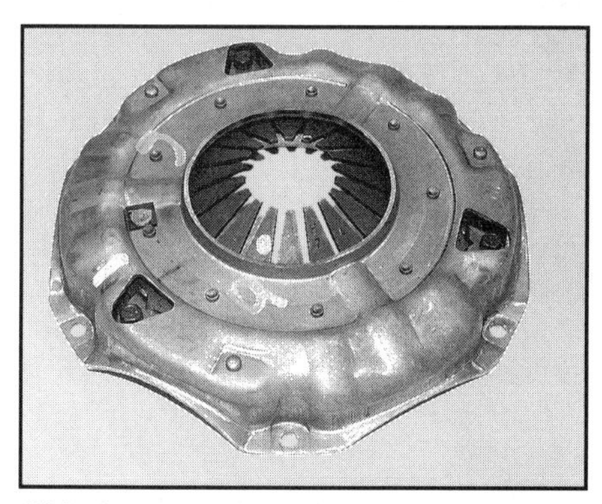

GM heavy duty 11-inch diameter diaphragm clutch cover must be used with 14-inch diameter flywheels and either the 11-inch, 10-tooth clutch disc PN 10148045 or the 11-inch 26-tooth clutch disc PN 10120997

As shown in the photos, late model crankshafts used in both the 300 HP and the 345 HP crate engines come with the small 3.00-inch diameter crank flange bolt pattern (bottom photo). The top photo shows a large 3.58-inch diameter bolt pattern as used on all pre-1986 engines. Both the HO and the SP engine come equipped with the correct flexplate.

To obtain proper starter engagement with 14-inch flywheels use a Chevy starter motor PN 1108400 which has staggered mounting bolt holes (top left). Use starter motor PN 1108789 with straight aligned mounting bolt holes for the smaller 12-3/4-inch flywheels and flexplates.

Chevy crate motors still accept the standard small block motor mount as used on all small blocks and most big blocks. If you are getting them at a parts store be sure to get the safety mount type with the built-in limiter that prevents separation if the rubber deteriorates.

350 SP 300 HP Engine Assembly

The special Performance 300hp, 350cid, PN 12355345, ID Code SP is designed for street use in 1968 and earlier cars or trucks. This 300hp special performance 350 GM engine assembly can also be used in any year off-road vehicle. Its 9.1:1 compression ratio delivers strong performance, and it can physically replace any small block engine from the 265 to the 400. The block is the same 4-bolt main, 1-piece rear main unit used for the High Output 350 smallblock. This engine uses non-swirl, 64cc cast iron cylinder heads and it includes a chrome timing cover and rocker covers. You must supply your own intake manifold, distributor, flywheel, balancer, water pump and exhaust manifolds. It requires a counterweighted flywheel PN 10105832, 14088646 or 14088650, a flex plate ('86 and later V8 PN 10128412 or 10128413), and the following balancers: PN 6272221 (6.75-inch), or PN 6272224 (8.00-inch) are recommended.

Keep in mind that you have a broad range of choices when it comes to intake and exhaust systems. While most people install the HO 350 engine as it comes from the factory, others have chosen aftermarket intakes and even electronic fuel injection. Both the HO 350 and the SP 350 are solid performance engines suitable for street rods, high performance street machines and even towing applications. In most cases your engine swap will not be smog legal unless approved by a referee station, or as specified in the HO 350 Camaro Coversion Package. Both engines will run best on premium unleaded fuel and you will obtain maximum performance if you incorporate a free flowing exhaust system, moderate rear axle gearing (3.23 to 3.73 ratios) and a stock or mild torque converter with a stall speed no higher than 2800 to 3000 rpm.

350 LT-1, 300 HP Engine Assembly

More compact size and new performance updates found in Chevrolet's relatively new LT-1 V8 have made it a favorite with engine swappers and street rodders. The LT1 has a compression ratio of 10.2:1 and it redlines at 5700 rpm. While not technically a crate motor, it can be purchased as a complete engine assembly and installed with the appropriate hardware and electronic controls. This second generation small block V8 incorporates significant advances in induction, exhaust, ignition and cooling. These changes required some departure from the small block's famous interchangeability, but the new LT1 is still a small block as we have always known them.

Reverse flow cooling has been designed into the LT1 and the front mounted water pump, camshaft drive and optical distributor are the most obvious changes. The engine also incorporates a short runner EFI intake manifold, hydraulic roller camshaft and high flow aluminum cylinder heads with heart-shaped combustion chambers. The front mounted distributor is driven by a short shaft from the camshaft sprocket. Inside the sealed housing a shutter wheel and optical sensor are used to provide the computer information about crankshaft position. This information is processed by the computer to determine optimum spark timing.

The reverse flow cooling system is designed to route water through the cylinder heads first and then into the cylinder block before returning to the water pump. Vents at the rear of the cylinder head prevent the formation of steam pockets and promote improved

300 HP LT1 350 ID Code LT1

SPEC SHEET:

GENERAL:	
Part Number	**10210312**
Displacement	350-cid, 5.7L
Bore & Stroke	4.00 x 3.48
Compression	10.2:1
Rated Power	300 @ 5000 rpm
Rated Torque	340 lb-ft. @ 4000
Fuel	92 octane, unleaded premium
CAMSHAFT:	
Part Number	**N/A**
Type	Hydraulic Roller
Valve Lift	.447/.459-inch
.050 Duration	
Int/exh lobe sep.	N/A
CYLINDER HEAD: (low port)	
Part Number	**1**
Material	Aluminum
Chamber Vol.	
Valve Material	Steel
Int. Valve Dia.	1.94
Exh. Valve Dia.	1.50
Rocker Ratio	1.5
Valve Springs	Standard
IGNITION:	
Distributor	Front-mounted optical drive

330 HP LT4 350 ID Code LT4

SPEC SHEET:

GENERAL:	
Part Number	**12551183**
Displacement	350-cid, 5.7L
Bore & Stroke	4.00 x 3.48
Compression	10.8:1
Rated Power	330 @ 5800 rpm
Rated Torque	340 lb-ft. @ 4500
Fuel	92 octane, unleaded premium
CAMSHAFT:	
Part Number	**N/A**
Type	Hydraulic Roller
Valve Lift	.476/.480-inch
.050 Duration	203°/210°
Int/exh lobe sep.	N/A
CYLINDER HEAD: (high port)	
Part Number	**1**
Material	Aluminum
Chamber Vol.	
Valve Material	Steel
Int. Valve Dia.	2.00
Exh. Valve Dia.	1.55
Rocker Ratio	1.6
Valve Springs	High Pressure
Intake Manifold	Short Runner
IGNITION:	
Distributor	Front-mounted optical drive

Factory Flywheels And Flexplates

The following chart includes part numbers and information about late model and early model flywheels and flexplates so you can be sure of selecting the correct one for your application. All late crate motors should use 1986 and later flywheels and flexplates, however an old replacement V8 or Target Master engine will use the earlier pieces.

Chevrolet Smallblock V8 and V6/90° Manual Transmission Flywheels

Part Number	Outside Diameter (inches)	Year	Crank Flange Bolt pattern	Clutch Diameter (inches)	Starter Ring Gear Teeth	Notes
14085720	12-3/4	1955-85	3.58	10.4	153	Weight approximately 15 pounds. This GM nodular iron flywheel was designed for 2-piece crank seals.
3986394	14	1970-80	3.58	11	168	This flywheel is designed for the 400ci smallblock V8 and is externally balanced. Compatlble with balancer PN 6272225.
3991469	14	1955-85	3.58	10.4, 11	168	For 2-piece seal cranks.
10105832	14	1986-up	3.00	11, 11.85	168	For 1-piece seal cranks.
14088646	12-3/4	1986-up	3.00	10.4, 11	153	Weight approximately 16 pounds. A lightweight nodular iron wheel for 1-piece seal cranks.
14088650	12-3/4	1986-up	3.00	10.4	153	Production-weight flywheel for 1-piece seal cranks.

Chevrolet Smallblock V8 and V6/90° Automatic Transmission Flexplates

Part Number	Outside Diameter (Inches)	Year	Crank Flange Bolt Pattern	Converter Bolt Pattern* (Inches)	Starter Ring Gear Teeth	Notes
471598	14	1970-85	3.58	10.75, 11.50	168	Use with crankshafts that do not require external balancing and have the early 2-piece rear seal.
471578	14	1970-80	3.58	10.75, 11.50	168	Use on externally-balanced 400cid engines. Compatible with balancer PN 6272225.
471529	12-3/4	1969-85	3.58	9.75, 10.75	153	Use with crankshafts that do not require external balancing and have the early 2-piece rear seal.
10128412	12-3/4	1986-up	3.00	10.75	153	Use on externally-balanced late-model cranks designed for the 1-piece rear. This is the flexplate specified for the 350HO engine.
10128413	14	1986-up	3.00	11.50	168	A heavy-dutly flexplate for externally-balanced late-model cranks designed for the 1-piece rear.
10128414	14	1986-up	3.00	10.75, 11.50	168	Use on externally-balanced late-model cranks designed for the 1-piece rear.

*NOTE: Torque converters for TurboHydro 350 and TurboHydro 400 transmissions are manufactured with both 10.75- and 11.5-inch converter bolt patterns. To identify the proper pattern, measure the distance from the converter centerline to an attaching bolt hole, then multiply by two.

cooling due to full contact between the liquid coolant and the water jacket. The water pump incorporates a two-way thermostat which mixes hot and cold coolant to reduce the thermal shock of cold coolant entering the engine. This improvement is not necessarily a major benefit to a crate motor application, but it is mentioned in passing so you will have a good understanding of the amount of new engineering that went into these late model engines.

350 LT-4, 330 HP Engine Assembly

Available as a complete stock replacement engine, the LT4 is the 1996 version of the LT1 with various revisions to increase power and reliability (see Spec Sheet on page 18). This 330 horsepower engine is installed as an option in 1996 Corvettes. It can be purchased separately under PN 12551183. Primary differences include lightweight hollow-stem valves with 2.00-inch diameter intakes and 1.55-inch diameter exhaust valves. The compression ratio has been increased from 10.2 to 10.8 and the hydraulic roller camshaft features a more aggressive profile with 203/210 degrees duration and .476/.480-inch valve lift, and for the first time, a production small block is equipped with 1.6 ratio rocker arms.

Still unavailable, but coming soon, the LT4 based H.O.T. high performance crate engine promises to deliver all the power and torque you could ask for in a high tech package designed to accomodate carburetion instead of EFI.

The LT4 requires a 1996 Electronic Control Module and wiring harness. As of early 1996, no aftermarket harnesses or ECMs are available. ECM's for these engines incorporate the new OBD II engine monitoring and on-board diagnostics electronics which makes it more difficult to reprogram them. However there is no reason why you couldn't make the engine run with an earlier computer and revised programing if you have access to the equipment and someone to program the electronics. Otherwise, you may want to consider the new single 4-barrel carburetor style H.O.T. LT4 crate motor recently developed by GM.

350 H.O.T. 421 HP Engine Assembly

At the time this book went to press, the LT4 based H.O.T. Chevy crate motor had not yet been approved for production, primarily because the LT4 engine is in short supply. Nevertheless all indications are that it will eventually be released to accomodate the demand for simplified, electronics free, high tech late model power plants.

The new crate motor is designed around the very capable parameters of the 330 HP LT4 engine. It delivers 421 horsepower at 6250 rpm 415 lbs-ft. of righteous torque at 4250 rpm. The short block for this new engine is identical to the LT4 short block, hence all the power gains have been accomplished with well designed bolt-on parts specified by GM engineers. This means that future rodders will be able to upgrade existing LT4s that show up in the scrap yard by applying the components listed here. The camshaft, carburetor and intake manifold package add more than 90 horsepower and 75 lbs-ft. of torque to the basic LT4 engine.

The new single four barrel intake manifold PN 24502574 is designed to accept both standard flange and spread-bore carburetors. And since the Gen II small block uses reverse flow cooling, a water crossover and thermostat housing are unnecessary. To accomodate the needs of all hot rodders, the manifold is machined to accept a standard rear mounted distributor.

The user has a choice of conventional ignition or the front mounted "Opti-spark" distributor. With "Opti-spark" the spark is triggered by a crankshaft-position sensor mounted on the aluminum front cover. This same sensor provides input for the OBD II diagnostics system on production engines. When this engine is released, a dedicated ECM and special wiring harness will be available at the same time to support the electronics if desired.

The H.O.T. crate motor takes maximum advantage of the LT4's high flow aluminum cylinder heads and lightweight sodium cooled valves. Lighter valve weight permits a higher 6300 rpm redline, and .100-inch taller intake ports achieve greater aiflow with an aggressive new camshaft PN 24502586 and 1.6

421 HP LT4 350 ID Code H.O.T.

SPEC SHEET:

GENERAL:

Part Number	N/A
Displacement	350-cid, 5.7L
Bore & Stroke	4.00 x 3.48
Compression	10.2:1
Rated Power	421 @ 6250 rpm
Rated Torque	415 lb-ft. @ 4250
Fuel	92 octane, unleaded premium

CAMSHAFT:

Part Number	PN 24502586
Type	Hydraulic Roller
Valve Lift	.525/.525-inch
.050 Duration	218°/228°

CYLINDER HEAD:

Part Number	N/A
Material	Aluminum
Valve Material	Steel
Int. Valve Dia.	1.94
Exh. Valve Dia.	1.50
Rocker Ratio	1.6:1
Intake Manifold	PN 24502574

IGNITION:

Distributor	Front-mounted optical drive

ratio rocker arms. The H.O.T. camshaft delivers .525-inch lift and 218/228 degrees of duration with the high lift rockers (.492-inch lift with standard 1.5:1 rockers). This camshaft still incorporates a standard type fuel pump eccentric lobe compatible with 1987 and later engine blocks.

This engine is the strongest small block crate motor ever conceived by Chevrolet. It delivers 76 more horspower than the ZZ3 HO crate motor and revs a thousand rpm higher. Torque is significantly greater and the broad overall operating range is perfect for high powered street machines. This is a sophisticated high tech nineties style engine that doesn't leave the average rodder in the dark when it comes to application and service. It offers plenty of high tech improvements for performance and reliability and small block fans are sure to love its power and versatility.

The H.O.T. crate engine offers a higher overall operating range thanks to lightweight hollow-stem valves filled with a sodium-potassium medium to promote valve and valve seat cooling. The valves measure 2.00-inch on the intake side and 1.55-inch on the exhaust side.

LT4 based aluminum cylinder heads have intake ports .100-inch taller than their LT1 counterparts. Recontoured exhaust ports also offer higher flow rates than the heads currently available on the H.O. 350 crate motor.

Much of the high tech value of the LT4 based H.O.T. crate motor is located on the front of the engine. In addition to the reverse flow collant water pump and front-mounted ignition system, the engine uses the crankshaft position sensor of it's on-board diagnostic system to trigger the ignition. As shown in the photo, the high voltage coil and module are mounted to the front of the right-side aluminum cylinder head.

New GEN VI version of the 454 and 502 big block are on the way in mid-1996. The changes are minor, but significant in the improvement they impart to the new engines.

GEN VI 454 & 502 Big Block Engines

Big block fans need not despair when searching for solid big block performance direct from GM. The current GEN V iterations of the 454 and 502 crate engines are quietly being upgraded to GEN VI status with a variety of modifications designed to provide even more performance advantages.

When GEN V engines were first introduced GM experienced temporary brain fade by neglecting to include important features such as a mechanical fuel pump mounting boss, hydraulic roller camshaft, a clutch linkage pivot stud mounting boss, oil cooler plumbing ports and block/cylinder head compatibility with earlier Mark IV cylinder heads.

These problems have all be fixed on GEN VI big blocks. A new composite front cover attaches with six bolts instead of ten. It addresses oil leakage with 0-ring sealed, torque-limiting shouldered mounting bolts. Another departure is the use of a timing notch on the front cover instead of the more familiar timing tab. A new heavy duty harmonic balancer marked in two degree increments is also incorporated to facilitate easier timing adjustments on the big block.

GEN VI engines will still feature high performance rectangular port cylinder heads with a high-rise aluminum intake manifold, a forged steel crankshaft, heavy duty connecting rods and pistons. The changes applied to the new GEN VI are considered normal evolutionary upgrades designed to make the engines efficient, more convenient and more reliable for all applications—sort of like adding more bulletproofing to the bulletproof big block.

The accompanying spec charts illustrate the primary content of the current GEN V engines and the part numbers are GEN V part numbers. Many of these engines still populate the parts system so make sure of what you are getting when you purchase a new big block. If the previously mentioned engine upgrades aren't of particular importance to your application, you can still use the GEN V version with great success.

If you are swapping to an earlier vehicle and need to retain a mechanical fuel pump, or you wish to install a good performing set of earlier Mark IV cylinder heads you will want the GEN VI engine. Additionally, the camshaft profile for the GEN VI has not yet been fixed. The hydraulic roller camshaft's faster valve action promises to deliver good improvements in low-speed and mid-range torque for street engines.

502 big blocks have found particular favor with high performance enthusiasts and the RV and towing crowd. Their superior torque and broad power band make them desirable for all kinds of replacement engine applications. Both the 454 and the 502 are built with heavy duty forged pistons that resist all the detonation you can throw at them with bad quality fuel. They will accept mild supercharging with no major changes and the bottom end is tough enough to withstand boost pressures up to 10 psi and a perhaps a bit more.

When installed in high performance street machines these engines offer all the power you could ask for. Factory EFI systems can be retro-fitted and you can really lean on these engines with out fear of damaging them. While separate marine big block versions are also available, may boat enthusiast have use 454 and 502 crate motors to significantly upgrade the performance of their existing marine packages.

425 HP HO 454

SPEC SHEET:

GENERAL:	
Part Number	**10185058**
Displacement	454-cid, 7.4L
Bore & Stroke	4.250 x 4.00
Compression	8.75:1
Rated Power	425 @ 5250 rpm
Rated Torque	467 @ 3500 rpm
Fuel	92 octane, unleaded premium
CAMSHAFT:	
Part Number	**10185060**
Type	Flat Tappet
Valve Lift	.510/.510-inch
.050 Duration	224°/224° @.050
Int/exh lobe sep.	116°
CYLINDER HEAD:	
Part Number	**14096801**
Material	Cast iron
Chamber Vol.	118cc, open
Valve Material	Steel
Int. Valve Dia.	2.19
Exh. Valve Dia.	1.88
Rocker Ratio	1.7
Intake Manifold	Aluminum Hi-Rise
SHORT BLOCK:	
Block	4-bolt mains
Crankshaft	1053 forged steel, 1- piece rear seal, PN14096983
Rods	4340 forged steel, 7/16 bolts, PN 10198822
Pistons	Forged Aluminum, PN 14097018

440 HP HO 502

SPEC SHEET:

GENERAL:	
Part Number	**10185085**
Displacement	502-cid, 8.2L
Bore & Stroke	4.466 x 4.00
Compression	8.75:1
Rated Power	440 @ 5250 rpm
Rated Torque	515 @ 3500 rpm
Fuel	92 octane, unleaded premium
CAMSHAFT:	
Part Number	**14096209**
Type	Flat Tappet
Valve Lift	.500/.500-inch
.050 Duration	220°/220° @.050
Int/exh lobe sep.	115°
CYLINDER HEAD:	
Part Number	**14096801**
Material	Cast iron
Chamber Vol.	118cc, open
Valve Material	Steel
Int. Valve Dia.	2.19
Exh. Valve Dia.	1.88
Rocker Ratio	1.7
Intake Manifold	Aluminum Hi-Rise
SHORT BLOCK:	
Block	4-bolt mains
Crankshaft	1053 forged steel, 1- piece rear seal, PN10183723
Rods	4340 forged steel, 7/16 bolts, PN 10198822
Pistons	Forged Aluminum PN 10198977

355ci, 325 HP Engine Assembly

This is a remanufactured small block engine assembly delivering 325 horsepower at 5250 rpm. It offers a 4-bolt main cylinder block, two piece rear main seal, cast iron cylinder heads with 76cc combustion chambers and 1.94/1.50-inch valves, hypereutectic pistons, with plasma moly rings, roller timing set, high volume oil pump, chrome front cover and chrome rocker covers. This is a good solid performance engine for street rods and engine replacement applications. It is available under PN 12360910.

325 horsepower 355 small block with 4-bolt mains offers a solid performance foundation for engine replacements and mild street machine projects.

All Chevrolet factory engines deliver good strong performance. Because they are factory assembled, they are subject to the same production inspections and test used on assembly line engines. This assures high quality and a high level of performance per the specifications of each individual engine package.

Small block and big block Chevy engines have become a staple of the high performance industry in America. This means that any Chevy crate motor you purchase can be easily upgraded or modified with well engineered high performance bolt-ons. Using crate motors as your foundation, you can build strong, powerful engines for any application imaginable.

Crate Motor Buyer's Guide

Ford Factory Engines

Photos courtesy of Ford SVO

5.0L GT-40 SVO Engine Assembly

Ford's 5.0L GT-40 SVO long block assembly is a solid high performance value rated at 285 horsepower with a 780 CFM Holley four barrel, dual plane intake and factory shorty headers. It is well suited for high performance applications thanks to the use of SVO GT-40 high-flow cast iron cylinder heads , GT-40 valve train and a SVO hydraulic roller cam.

The 9:1 compression 5.0L GT-40 SVO engine is built at Ford's Cleveland engine plant on the same production line as Mustang 5.0L engines. It is a bolt-in assembly for all late model Mustang and Fox chassis cars plus F-150 pickups. The short block assembly includes the block, crankshaft, connecting rods, pistons, timing chain, standard rocker arms, roller tappets, push rods, oil pump and pickup, oil pan, valve covers, front covers, water pump, damper, manual transmission flywheel, spark plugs and related long block engine parts. These engines come with the "short" front cover and water pump. Your original front cover and water pump will be necessary for use in other applications.

The significant power increase over a stock 5.0L engine is achieved with the use of high-flow SVO cast iron performance cylinder heads, PN M-6049-L302. These revised castings flow considerably more air than the standard production 5.0L iron cylinder head. SVO engineers also added the SVO GT-40 Valve Train Kit, PN M-6090-L302 which includes valve springs, retainers, keepers, seals and premium stainless steel valves with undercut stems and swirl polished heads. Valve diameters are 1.84-inch and 1.54-inch for the intake and exhaust valves respectively.

The final link is the power department is the use of SVO high performance hydraulic roller cam PN M-6250-B303. It offers a valve lift of 0.480-inch with 224° duration at .050-

Ford's 5.0L GT-40 SVO engine assembly is Street Legal with the installation of a stock camshaft and other emission-related components.

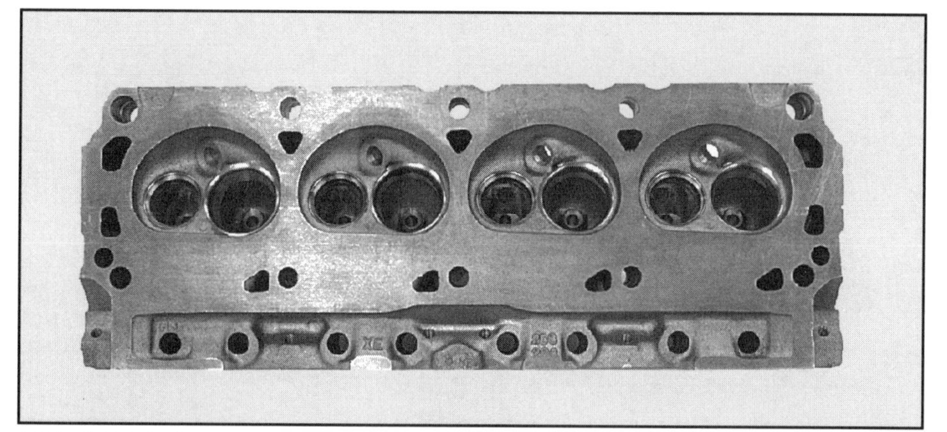

5.0L GT-40 SVO engines are equipped with cast iron GT-40 cylinder heads that flow considerably better than stock 5.0L iron heads and still retain full exhaust heat and EGR provisions for emissions and good street drivability.

inch lift. This cam retains good low end power characteristics while increasing power significantly above 4000 rpm. The 285 horsepower rating is a relatively conservative number and most engines actually deliver closer to 300 horsepower.

Ford's 5.0L GT-40 SVO engine assembly is street legal with the installation of a stock camshaft and other emission-related components. This engine was developed for use with a carburetor or electronic fuel injection depending on the application. The quoted horsepower ratings were established with a 780 CFM Holley four barrel carburetor, a high rise dual plane intake, PN M-9424-A321, and factory shorty exhaust headers.

Cobra EFI Manifold Assembly PN M-9424-D51

Fuel injected versions may experience a shift in the power curve due to the longer, individual runner lengths of the Cobra EFI intake manifold, M-9424-D51, and the GT-40 EFI manifold, M-6001-A50. Factory EFI engines tend to generate stronger low end torque with slightly less power at high rpm. These intakes overcome this for the most part and make primo induction systems for almost any application.

This engine is particularly

5.0L GT-40 SVO

SPEC SHEET:

GENERAL:

Part Number	**M-6007-A50**
Displacement	302-cid, 5.0L
Bore & Stroke	4.00 x 3.00
Compression	9.0:1
Rated Power	285 @ 5200 rpm
Rated Torque	327 @ 4000 rpm

CAMSHAFT:

Part Number	**M-6250-B303**
Type	Hydraulic Roller
Valve Lift	0.480-inch
.050 Duration	224°
Int/exh lobe sep.	112°
Int. Centerline	107°

CYLINDER HEAD:

Part Number	**M-6049-L302**
Material	Cast iron
Chamber Vol.	65.5cc
Int. Valve Dia.	1.84
Exh. Valve Dia.	1.540

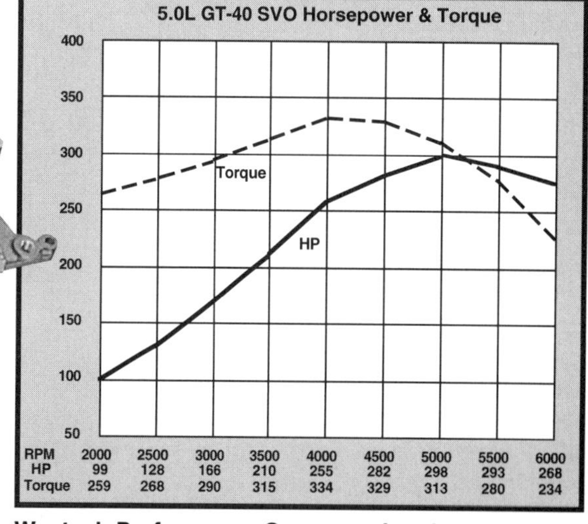

RPM	2000	2500	3000	3500	4000	4500	5000	5500	6000
HP	99	128	166	210	255	282	298	293	268
Torque	259	268	290	315	334	329	313	280	234

Westech Performance Group engine dyno test with factory EFI, SVO 55mm air meter, 19 lb./hr. fuel injectors and shorty headers.

What You Don't Get

- **Induction system**
- **Distributor**
- **Ignition Module**
- **Spark plug wires**
- **Headers**
- **Accessory drives**
- **Motor mounts**
- **Flywheel**
- **Starter**
- **Air cleaner**

5.0L GT-40 Installation Tips

OIL PAN
Check for interference if installing in other than a Fox chassis

FRONT COVER
Use correct design, with or without fuel pump mounting, and with correct water pump compatibility

FLYWHEEL/DAMPER
Use proper type, 157 or 164 tooth flywheel and 3- or 4-bolt with 50 oz. unbalance production damper

DISTRIBUTOR
Use type compatible with roller cam
Use 351W firing order 1-3-7-2-6-5-4-8

well suited for replacement use on high mileage cars and trucks. Mustang owners will acquire a solid performance gain simply by installing the 5.0L GT-40 engine assembly and reusing their existing intake and exhaust systems.

The "Cobra" EFI intake manifold used as original equipment on Cobra R Mustangs is a good choice for adding EFI to the 5.0L GT-40 engine. It is about half the price of the standard GT-40 intake system and will support up to 400 HP. The upper half has staggered ports that only match the lower half of the GT-40 intake manifold. It cannot be installed on a stock production 5.0L "base" or GT lower intake manifold with the inline port configuration. Both the"Cobra" and the GT-40 intake manifolds can be used with stock production fuel rails, sensors and wiring harnesses to ensure maximum compatibility with all factory EFI components. The accompanying engine dyno test results were accomplished using only the production intake manifold, a 55mm air meter and factory shorty headers.

The strongest factory 302 crate motor is the 5.0L GT-40 assembly with aluminum cylinder heads. It is rated at 320 HP when used with the SVO GT-40 EFI induction kit, factory headers and a 65mm throttle body assembly.

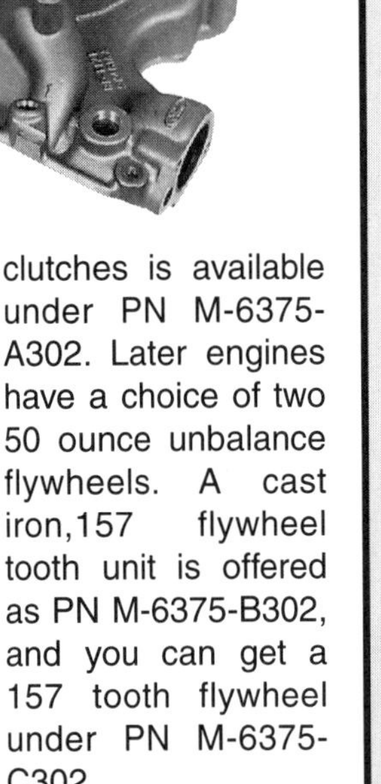

GT-40 EFI Manifold Assembly PN-M-6001-A50

If you're installing any of the 5.0L or 351 V8 engines as replacement engines in late model applications you must use the production harmonic balancer which has a built-in extension to allow proper alignment of the serpentine belt accessory drive system. The Motorsport dampers PN M-6316-A50 and M6316-C351 will not align with serpentine drive belts unless you also use a special crankshaft pulley adapter from Advanced Engineering (909) 930-9852 which spaces the pulley away from the damper for proper alignment.

To determine the correct flywheel for manual transmission applications keep in mind that pre-1981 302 engines use a 28.2 ounce unbalance flywheel while all later 302 engines use a 50 ounce unbalance. A 28.2 ounce unbalance, billet steel 157 tooth flywheel for 10.5-inch HD clutch assemblies and 10.5-inch long style clutches is available under PN M-6375-A302. Later engines have a choice of two 50 ounce unbalance flywheels. A cast iron,157 flywheel tooth unit is offered as PN M-6375-B302, and you can get a 157 tooth flywheel under PN M-6375-C302.

Metric fasteners and dowel pins supplied with these flywheels, MUST be used with the 10.5-inch clutch assembly. 1986 and later Mustang/Capri 302 vehicles with T-5 transmissions have standard duty 10.5-inch clutches. You

CRATE MOTOR Performance Tips & Tricks

GT-40 Aluminum Heads PN M-6049-Y303

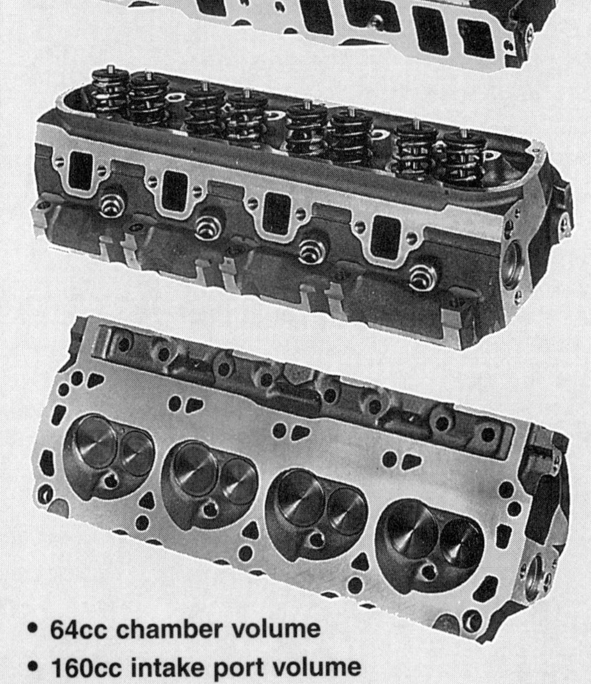

- 64cc chamber volume
- 160cc intake port volume
- 53cc exhaust port volume
- 1.94/1.54 intake/exhaust valves
- 356-T6 alloy casting
- All SVO valve train
- 22 lbs. each, 25 lbs. lighter than stock
- 40 HP increase over stock 5.0L heads
- Bare Casting, PN M-6049-Y302
- Spark Plugs, AGSF-32C

The 285 HP "Cobra" 302 shortblock assembly is a good starting point for those who are looking for a replacement engine assembly, and are able to incorporate all of their existing ignition, exhaust and induction hardware. It is available under PN M-6009-B50. A single plane high rise "Victor Jr." intake PN M-9424-D302 makes a good choice for hot street use with the "Cobra" short block and aluminum or iron GT-40 cylinder heads.

can use the HD clutch PN M-7563-A302 or the "King Cobra" clutch PN M-7563-C302 on the production flywheel with the production fasteners and dowel pins. The "King Cobra" clutch requires about 10% less pedal effort and is preferred for street applications.

A high performance Clutch Linkage Upgrade Kit PN M-7553-A302 is recommend for all performance applications. The correct replacement throwout bearing for all 1979 and later Mustang V8 models with manual transmissions and production or Motorsport HD clutch assemblies is D9ZZ-7548-A. All of these components are currently available from any local Ford or Lincoln-Mercury dealer. They are good insurance against future problems which might occur once you have the car running.

5.0L GT-40 SVO Engine Assembly w/Aluminum Heads

The aluminum cylinder head version of the 5.0L GT-40 engine assembly is essentially the same engine as the previously described 5.0L GT-40 SVO engine assembly. The short block is identical, including the same B303 hydraulic roller camshaft, and the compression ratio remains the same at 9:1.

The only difference is the addition of SVO's GT-40 aluminum cylinder heads PN M-6049-Y303. These heads feature "turbo swirl" combustion chambers with enlarged intake and exhaust ports and a bigger 1.94-inch diameter intake valve (see sidebar).

For optimum results you'll want to use the aluminum GT-40 intake mani-

5.0L GT-40 SVO Installation Tips

OIL PAN
Check for interference if installing in other than a Fox chassis

FRONT COVER
Use correct design, with or without fuel pump mounting, and with correct water pump compatibility

FLYWHEEL/DAMPER
Use proper type, 157 or 164 tooth flywheel and 3- or 4-bolt with 50 oz. unbalance production damper

DISTRIBUTOR
Use type compatible with roller cam
Use 351W firing order 1-3-7-2-6-5-4-8

5.0L 302 Engine

5.0L Mustang "94-'95 HO EFI *New Engine Assembly*

This engine is no longer produced, but it is still available while supplies last. It is a great choice for most engine swaps and you can get the whole package for less than $3000. The following features are common to this engine.

- **302 cubic inch, 240 HP**
- **9:1 Compresion ratio**
- **SVO E303 Hi-Po roller cam**
- **60mm throttle body**
- **Stainless headers**
- **Complete fuel injection**
- **Distributor**
- **Fuel rails**
- **Injectors**
- **Injector harness**
- **Short front cover & water pump**
- **Uses serpentine pulley drive**
- **New current production 5.0L engine with new block, crank, rods, pistons, timing chain, rocker arms, roller tappets, push rods, oil pump, and pickup, oil pan, valve covers, front covers, water pump, damper, flywheel and spark plugs**
- **Fits early Mustangs, trucks, Fox bodied cars street rods and kits cars with no modifications**
- **Street legal when installed with related emissions components**
- **The ideal engine assembly for all stock replacement applications and most high performance engine swap projects.**

fold, PN M-6001-A50. It features long, 1.65-inch diameter staggered port runners and is rated up to 425 horsepower. Like the Cobra intake, the GT-40 intake manifold accepts the stock fuel rail, sensors and wiring harness, and it is a bolt-on in most applications which are already fuel injected. For previously carburetted applications, all the components of a stock EFI system will be necessary to effect a complete and successful EFI conversion.

The production 5.0L 302 "Cobra" short block assembly PN M-6009-B50 is an attractive alternative for those who may already have a good set of cylinder heads, or as a replacement for a worn engine. It is a 2-bolt main assembly with hypereutectic pistons, 9:1 compression and a high torque roller cam featuring 270° duration and 0.480-inch lift. It is rated at 285 HP when teamed with the Motorsport induction, GT-40 heads, headers and a 65mm throttle body assembly. The Cobra short block is essentially the same assembly used to make both 5.0L GT-40 long block assembles.

5.0L GT-40 SVO with Aluminum Heads

SPEC SHEET:

GENERAL:

Part Number	**M-6007-B50**
Displacement	302-cid, 5.0L
Bore & Stroke	4.00 x 3.00
Compression	9.0:1
Rated Power	320 @ 5200 rpm
Rated Torque	327 @ 4000 rpm

CAMSHAFT:

Part Number	**M-6250-B303**
Type	Hydraulic Roller
Valve Lift	0.480-inch
.050 Duration	224°
Int/exh lobe sep.	112°
Int. Centerline	107°

CYLINDER HEAD:

Part Number	**M-6049-Y302**
Material	Aluminum
Chamber Vol.	64cc
Valve Material	Stainless steel
Int. Valve Dia.	1.94
Exh. Valve Dia.	1.540

A top of the line 351 HO SVO engine assembly offers plenty of bang for the buck with a 10:1 compression ratio and 385 horsepower. This engine comes with aluminum heads, roller rockers, flat tappet camshaft and SVO "Victor Jr." single plane intake manifold. Note that the engine does not include an air cleaner, carburetor, wiring or exhaust headers. This engine comes with no warranty.

351 HO SVO Engine Assembly

One of the hottest choices for street and bracket racing applications is the 351 HO SVO engine assembly. This engine is suitable for late-model short track applications, modified competition or marine use. It has a 10:1 compression ratio with hypereutectic pistons and can be run on high quality pump gas with appropriate tuning. Ford rates the engine at 385 horsepower with a 780 CFM Holley carburetor and dyno headers. Power peaks at 5750 RPM with 377 lbs-ft of torque at 4500 RPM. Each engine is test fired and ready to install. The 351 HO SVO fits all Mustang and Fox chassis cars and can be adapted to earlier applications. F-series trucks will accept this engine with a different oil pan.

This very high performance engine is based on a Ford 351W 2-bolt main marine cylinder block with a nodular iron crank, 10:1 high silicon pistons and HD marine/truck connecting rods. These connecting rods are specially prepared with spot-faced seats for the HD 3/8-inch bolts.

The cylinder heads are SVO Windsor high flow aluminum units PN M-6049-J302 with SVO premium stainless steel valves featuring 1.94-inch diameter intakes and 1.60-inch diameter exhaust valves. Minor porting and a 3-angle valve job highlight the cylinder head package along with

Suggested Front Cover Combinations For Correct 351W V-belt Installation

Application	Water Pump	Cover	Damper	Pointer
'69 & earlier, 3-bolt crank pulley with passenger side water inlet	D3UZ-8501-A	E7AZ-6019-A*	C9OZ-6316-A**	C8AZ-6023-A**
'70 & later 4-bolt crank pulley, driver's side water pump inlet	D6ZZ-8501-B	E7AZ-6019-A*	Use Crate Engine Damper	C9TZ6023-A

* Works with serpentine belt drives when running a mechanical fuel pump. Grind away crank trigger mount as required for timing pointer clearance.

** No longer available. Must be found in salveage yards.

SVO aluminum roller rockers PN M-6564-A351 (1.60 ratio), SVO dual valve springs and SVO step-up pushrod guide plates PN M-6566-B302. A dual mounting bolt pattern allows the use of stock exhaust manifolds or headers with up to 1$^3/_4$-inch diameter primary tubes.

A very strong high performance hydraulic flat tappet style camshaft supports this engine package. It is a dual pattern grind with a maximum valve lift of .520-inch intake and .538-inch on the exhaust. The duration at .050-inch valve lift is 236 degrees on the intake and 246 degrees on the exhaust. Factory technicians install the cam 2 degrees retarded using an SVO multi-index true roller timing chain PN M-6268-A302. This cam uses SVO anti-pump up hydraulic valve lifters PN M-6500-A301. The engine is equipped with extra tall SVO aluminum valve covers PN M-6582-E302 which clear roller rockers.

The induction system incorporates a "Victor Jr." high rise, single plane aluminum intake manifold PN M-9424-V351 to optimize high RPM power. A 750 or 780 CFM Holley 4-barrel is recommended for optimum performance. To ensure optimum power with this induction system Ford supplies a OEM 351W electronic distributor PN E2AZ-12127-E coupled with a production ignition coil PN E8TF-12029-BA. An ignition module is not included, but Duraspark II, Extra Performance or Ultra CD ignition systems are recommended.

To ensure top durability, Ford combines a production oil pump and a HD chrome moly drive shaft PN M-6605-A341. The dual sump 5-quart oil pan clears the crossmember on all Fox bodied cars, but it will have to be changed if you are using the engine in an earlier application. These applications can use a front sump pan 351W oil pan PNMD5OZ-6675-A and pump pickup PN(D7AZ-6622-A. The late model one-piece rear main seal does not interfere with oil pan interchange in this application.

The 351 HO SVO engine comes with a steel billet 157 tooth flywheel PN M6375-A302 which accepts 10.5-inch Long or diaphragm clutches. This flywheel uses 28.2 ounces of unbalance, the same as most 1980 and earlier 289/302 engines.The bellhousing and flywheel size must match in order to mount the starter correctly. The crankshaft damper is a 4-bolt design which fits all 1970 4-bolt crank pulleys. It is designed for late model serpentine belt accessory drives and a electric fuel pump.

The 351 GT-40 EFI engine assembly is a complete high performance crate engine capable of delivering 310 HP with the SVO cam, PN M-6250-B351. It fits all Mustang and Fox bodied cars and can be adapted to earlier chassis. It features GT-40 cast iron heads, GT-40 upper and lower intake manifold and a 65mm throttle body.

Late model 351W blocks no longer have a clutch-shaft pivot-ball hole. If you are installing the engine in an earlier car that requires a bell-crank mechanical linkage, you can use a pivot-ball mounting adapter from Total Performance (810/468-3673), or convert to hydraulic linkage by using McLeod Industries (714/630-2764) integral hydraulic throwout bearing. Kaufmann Products (310/803-5531) also offers a cable clutch linkage conversion kit.

Because the 351 W block is approximately 1 inch taller and 1.1-inch wider than the 289/302 engine, 351 accessory brackets are required when changing over from the smaller engines. Ford Motorsports PN M-8511-A351 is the correct power steering bracket while PN M-8511-B351 is a combination

351 HO SVO

SPEC SHEET:

GENERAL:	
Part Number	**M-6007-A351**
Displacement	351-cid, 5.8L
Bore & Stroke	4.00 x 3.50
Compression	10.0:1
Rated Power	385 @ 5750 rpm
Rated Torque	377 @ 4500 rpm
CAMSHAFT:	
Part Number	**M-6250-A351**
Type	Hydraulic
Valve Lift	.520/.538 I/E
.050 Duration	236°/246° I/E
CYLINDER HEAD:	
Part Number	**M-6049-J302**
Material	Aluminum
Chamber Vol.	64cc
Valve Material	Stainless steel
Int. Valve Dia.	1.94
Exh. Valve Dia.	1.60

power steering and A/C bracket.

If a mechanical fuel pump and/or early style V-belt accessory drive are used, an early 351W front cover must be installed along with an inner and outer fuel pump eccentric on the front of the camshaft and a standard rotation water pump. Some applications may encounter differences in the crankshaft damper, water pump, accessory drive mounting brackets, clutch linkage, flywheel, flexplates and induction if you install electronic fuel injection in place of the carburetor. Ford Motorsport recommends calling their Technical Hotline (313) 337-1356 for assistance is solving any unusual installation and application problems.

The 351 GT-40 SVO Long block assembly is a hot engine with a budget price. It features a high performcne flat tappet camshaft, SVO GT-40 cast iron cylinder heads and appropriate components to produce 346 horsepower with Motorsport Induction and headers.

351GT-40 EFI Engine Assembly

This fuel injected engine is based on a production Ford truck 351W 2-bolt cylinder block with a nodular iron cast crankshaft, 8.8:1 hypereutectic pistons with plasma moly-filled rings and heavy duty Ford truck connecting rods. It offers 310 horsepower if the production camshaft is replaced with an SVO camshaft PN M6250-B351. This cam makes 340 lbs-ft of torque at 3200 rpm. Each engine is brand new, pre-fired and ready to install.

High flow cast iron GT-40 cylinder heads with 1.84/1.54-inch diameter valves are incorporated with exhaust seat inserts to prevent seat recession. The camshaft is a high torque unit featuring 260 degrees intake duration and 274 degrees on the exhaust side. Valve lift is .416/.444. This combination uses standard 1.60 ratio rockers and a production roller timing chain. A production oil pump and late-model truck 6-quart oil pan are standard. A water to oil heat exchanger is included

351 GT-40 SVO

SPEC SHEET:

GENERAL:	
Part Number	**M-6007-A351**
Displacement	351-cid, 5.8L
Bore & Stroke	4.00 x 3.50
Compression	10.0:1
Rated Power	385 @ 5750 rpm
Rated Torque	377 @ 4500 rpm
CAMSHAFT:	
Part Number	**M-6250-A351**
Type	Hydraulic
Valve Lift	.520/.538 I/E
.050 Duration	236°/246° I/E
CYLINDER HEAD:	
Part Number	**M-6049-J302**
Material	Aluminum
Chamber Vol.	64cc
Valve Material	Stainless steel
Int. Valve Dia.	1.94
Exh. Valve Dia.	1.60

The 351 GT-40 "Lightning" short block PN M-6009-B58 offers the same basic assembly in short block form for those who wish to furnish their own heads and accessories. It produces 310 horsepower with GT-40 heads and production cam. This assembly comes with a roller timing set, hypereutectic pistons, moly top rings and all new components.

to provide supplemental oil cooling.

The engine is complete with ignition system, stainless tubular headers, wiring harness and spark plug wires. It uses late model production 351W components that are compatible with serpentine belt drive systems equipped with electric fuel pumps. If a mechanical fuel pump or a V-belt drive system is desired, an early 351W front cover must be used with a standard rotation water pump. The camshaft must also be equipped with the inner and outer fuel pump eccentric. Minor differences in the crankshaft damper, water pump, accessory drive mounting brackets, clutch linkage, flywheel, flexplates and induction may require attention during installation of this engine.

The 351 GT-40 EFI engine is the perfect replacement engine for late model cars and trucks with well worn engines, or for early model vehicles where an engine swap is contemplated. With the use of a factory wiring harness and ECM control box you can make this engine perform well in any early model car. It offers good power and torque in stock form and a complete lineup of high performance factory and aftermaket parts are available for hopping it up.

351GT-40 SVO Long Block Engine Assembly

Ford calls the 351 GT-40 SVO engine the fast and easy way to add performance on a limited budget. This engine offers good performance to value ratio with 346 horsepower and 364 lbs-ft of torque available when teamed with a 750 CFM Holley carb and tubular headers. The compression ratio is 9:1 and the SVO cast iron heads will accept late model EFI intakes for those wishing to run fuel injection or those wishing to replace a tired EFI motor with a fresh long block assembly. Your choice of induction and ignition is left open. Ford recommends this engine for boosting performance in older cars, modified competition and marine applications. The engine fits all late model Mustangs and Fox bodied cars.

The short block is built around a proven Ford marine 351W 2-bolt main block equipped with a nodular iron crankshaft, 9.0:1 hypereutectic pistons and heavy duty marine/truck connecting rods. SVO high flow cast iron GT-40 cylinder heads PN M6049-L302 with premium 1.84/1.54-inch diameter stainless steel valves and hardened valve seats are standard.

A high performance SVO hydraulic flat tappet camshaft PN M-6049-L302 is part of the 351 GT-40 SVO package. It is a dual pattern grind offering .491/.509-inch lift with 220°/230° duration at .050-lift. The valve train uses

351 GT-40 EFI

SPEC SHEET:

GENERAL:

Part Number	**M-6007-L58**
Displacement	351-cid, 5.8L
Bore & Stroke	4.00 x 3.50
Compression	8.8:1
Induction	EFI
Rated Power	310 @ 5400 rpm
Rated Torque	340 @ 3200 rpm

CAMSHAFT:

Part Number	Production
Type	Hydraulic
Valve Lift	.414/.444 I/E
.050 Duration	260°/274° I/E

CYLINDER HEAD:

Part Number	**M-6049-L303**
Material	iron
Chamber Vol.	65.5cc
Valve Material	Steel
Int. Valve Dia.	1.84
Exh. Valve Dia.	1.54

CLUTCH & FLYWHEEL ASSEMBLIES

Part Number	Type	Diameter	No. of Teeth	Unbalance
M-7563-A302	Press. Plate	10.5		
M-7563-C302	Press. Plate	10.5		
M-7550-A302	Clutch Disc	10.5	10 spline	
M-6375-A302	Flywheel/billet		157	28.2 oz.
M-6375-B302	Flywheel/cast		157	50.0 oz.
M-6375-C302	Flywheel/billet		157	50.0 oz.

10.5-inch Cluth Notes

- **Vehicles equipped with a 10-inch clutch require a new flywheel. Metric fasteners & dowel pins furnished with these flywheels must be used with with the new 10.5-inch clutch. Use billet flywheel for high rpm applications.**
- **1986 and later Mustang/Capri 302 vehicles with T-5 transmissions already have standard duty 10.5-inch clutches. HD 10.5-inch clutches can be installed using the production flywheel, fasteners and dowel pins.**
- **Clutch linkage Upgrade kit PN M-7553-A302 recommended for all applications.**
- **1979 and later Mustang V8s with manual transmissions and Ford Motorsport or production clutch assemblies require clutch throw-out bearing D9ZZ-7548-A.**
- **"King Cobra" HD presure plate PN M-7563-C302 10.5-inch clutch pressure plate has the same torque capacity and nodular iron pressure plate as M-7563-A302. It requires approximately 10% less pedal effort due to a new, stronger cover and revised geometry. these units are interchangeable.**

standard 1.6 ratio rockers with the SVO valve train kit PN M-6090-L302 (springs and retainers). SVO "tall" Motorsport signature style valve covers are included.

The engine comes with a 164 tooth automatic transmission flexplate, but you can add the appropriate 50 ounce unbalance flywheel for manual transmissions (see accompanying chart). The dual-sump, 5-quart oil pan is a production unit that clears most Fox-body chassis, but may interfere on some cars. The SVO damper is the correct 4-bolt design to fit all 1970 and later 4-bolt crank pulleys.

All late model Ford crate engines are built from production parts so many of the installation particulars apply to all engines. The 351 GT-40 SVO engine requires the same installation hardware and modifications needed for both the 351 HO SVO and the 351GT-40 EFI engines.

The 460 "Cobra Jet" crate motor is a 560 horsepower big block Ford designed to satisfy even the most demanding crate motor application. Backet racers and Ford street enthusiasts can make good use of this high compression powerhouse, but it does require high quality fuel plus additives. It will also tear the tires off of anything you are likely to install it in.

460 "Cobra Jet" SVO Engine Assembly

The mother of all Ford big Blocks is this 460 cubic inch monster designed to put the ultimate grunt in any crate motor engine swap. This thumper delivers 560 horsepower at 6000 rpm and 535 lbs-ft of torque at 4750 rpm. With a compression ratio of 11.5:1, this is a seriously hot piece built for use in bracket racing, marine applications, short track, modified competition and hot pro street machines. Everything is included except carburetor, ignition, headers and accessory drives.

The short block features a nodular iron crank with HD forged rods, aluminum hypereutectic pistons and roller timing chain. The aluminum SVO 460 cylinder heads feature a high-flow Jon Kasse port design that complements the SVO Victor Jr. intake manifold PN M-9424-G429. This manifold is optimized for high end power when used with an 850 CFM Holley carburetor. These heads bolt directly onto all production 460 engines. They use production-type valves, springs, retainers and other valve train components. Combustion chamber volume is set at 72cc.

This engine uses late model production parts designed for serpentime belt accessory drive systems and a electric fuel pump.

Production belt drive systems use reverse rotation water pump PN F2TE-8501-AA. The front cover and water pump must be changed if a V-belt drive is desired. A fuel pump eccentric must also be added to the camshaft to operate a mechanical fuel pump. This engine will tear the tires off of anything you install it in.

460 "Cobra Jet"

SPEC SHEET:

GENERAL:		
Part Number	**M-6007-C460**	
Displacement	460-cid, 7.5L	
Bore & Stroke	4.36 x 3.85	
Compression	11.5:1	
Rated Power	560 @ 6000 rpm	
Rated Torque	340 @ 4750 rpm	
CAMSHAFT:		
Part Number	N/A	
Type	Roller	
Valve Lift	.614/.614	I/E
.050 Duration	244°/254°	I/E
CYLINDER HEAD:		
Part Number	**M-6049-A429**	
Material	Aluminum	
Chamber Vol.	72cc	
Valve Material	Steel	
Int. Valve Dia.	2.25	
Exh. Valve Dia.	1.75	
Rockers	Roller	
Rocker Ratio	1.73	

Photos courtesy of Mopar Performance

Magnum 300 Engine
300 HP 360 V8

Only two engines are available in the Mopar line, but their performance level is second to none. The base level engine is a 300 horsepower 360 cubic inch V8 built with all new factory parts. In keeping with Mopar tradition, it is a complete assembly from oil pan to intake manifold and it even includes a high performance electronic ignition package.

Built with a standard production cylinder block and cast crankshaft, it is also equipped with a smooth, torque producing hydraulic roller camshaft and swirl port iron cylinder heads. These cylinder heads are outfitted with 1.925-inch intake valves and 1.625-inch exhaust valves with high performance valve springs. An M1 aluminum dual plane intake manifold is standard and dyno testing has established a 300 horsepower level when equipped with a 750 CFM Holley four barrel and 1-5/8-inch diameter exhaust headers. This power is achieved at 4750 RPM, while peak torque is 375 lb-ft at 4000 rpm.. Over 300 lb-ft of torque is available from 3000 to 5000 rpm. These Magnum engine assemblies are shipped with new cast alu-

Magnum 300 Engine assembly measures 360 cubic inches and is designed to provide a powerful, convenient engine swap package for stock replacement applications or moderate performance street machines and towing packages. The engine is dressed out with black wrinkle finish Magnum valve covers and chrome accents.

Production forged connecting rods measure 6.123-inch center to center, and are fitted with high strength rod bolts. Factory cast aluminum pistons provide a nominal 9:1 compression ratio with the Magnum cylinder head.

Magnum 300 cast iron cylinder head is a free breathing piece with generous 1.925-inch intake valves and 1.625-inch exhaust valves. On the Magnum engine, these cylinder heads are fitted with high performance valve springs, retainers and keepers to ensure stable valve train operation.

Factory 3.58-inch stroke iron crankshaft is a solid piece that will take plenty of abuse. Main journals measure 2.81-inch with rod journals at 2.125-inch.

The 300 horsepower rating is achieved with a 750 CFM carburetor and 1-5/8-inch diameter headers. This dual plane aluminum intake manifold is responsible for delivering over 300 lb-ft. of torque from 3000 rpm to 5000 rpm and a torque peak of 375 at 4000 rpm.

Magnum 300
300 HP 360 V8

SPEC SHEET:

GENERAL:

Part Number	**P5249498**
Displacement	360-cid, 5.9L
Bore & Stroke	4.00 x 3.58
Compression	9.0:1
Rated Power	300 @ 4750 rpm
Rated Torque	375 @ 4000 rpm

CAMSHAFT:

Part Number	N/A
Type	Hydraulic Roller
Valve Lift	0.480-inch
.050 Duration	224°
Int/exh lobe sep.	112°
Int. Centerline	107°

CYLINDER HEAD:

Part Number	**P5249573**
Material	Cast iron
Valve Material	Steel
Int. Valve Dia.	1.925
Exh. Valve Dia.	1.625

The 380 horsepower Magnum 380 engine assembly is Mopar Performance's top of the line engine offering. Some aftermarket tests have shown this engine delivering closer to 400 horsepower, and the torque level is greater than 350 lbs-ft. from 2000 rpm to 5700 rpm.

Magnum 380
380 HP 360 V8

SPEC SHEET:

GENERAL:	
Part Number	**P5249499**
Displacement	360-cid, 5.9L
Bore & Stroke	4.00 x 3.58
Compression	9.0:1
Rated Power	380 @ 5300 rpm
Rated Torque	410 @ 4400 rpm
CAMSHAFT:	
Part Number	**P5249549**
Type	Hydraulic Roller
Valve Lift	0.501/0.513-inch
.050 Duration	288°/292°
Int/exh lobe sep.	112°
Int. Centerline	107°
CYLINDER HEAD:	
Part Number	**P5249573**
Material	Cast iron
Valve Material	Steel
Int. Valve Dia.	1.925
Exh. Valve Dia.	1.625
Valve Spring	P5249464

minum valve covers with a raised Magnum logo.

These engines are not generally legal for use on emission controlled vehicles in California and other states. Purchasers must sign a Purchase Agreement stating that the engine is for off-road use only. Of course many older Mopar vehicles that are not regulated are fair game for the easy power swap that Magnum engines provide.

Magnum 380 Engine 380 HP 360 V8

Previously available Mopar Commando engines have been replaced with the new Magnum 380. According to Mopar Performance, the magnum engine family is developed to a higher level than the Commando series even with modifications. This allows Magnum engine to show a larger power gain from similar modifications previously made to the Commando engines.

The Magnum 380 cylinder head is a iron swirl port design with smaller combustion chambers and a high compression ratio (nominally 9.0:1). Revised intake and exhaust port sizing flows more air without additional modification to the cylinder heads. The valves are 1.925-inch in diameter on the intake side and 1.625-inch on the exhaust side. High performance valve springs are standard.

A very strong .501"/.513" lift, 288°/292° duration hydraulic roller camshaft complements the new port flow capacity and a 4 barrel M1 single plane aluminum intake manifold is attached for optimum induction efficiency. This intake manifold is already equipped with built-in bosses over each runner for future port fuel injectors.

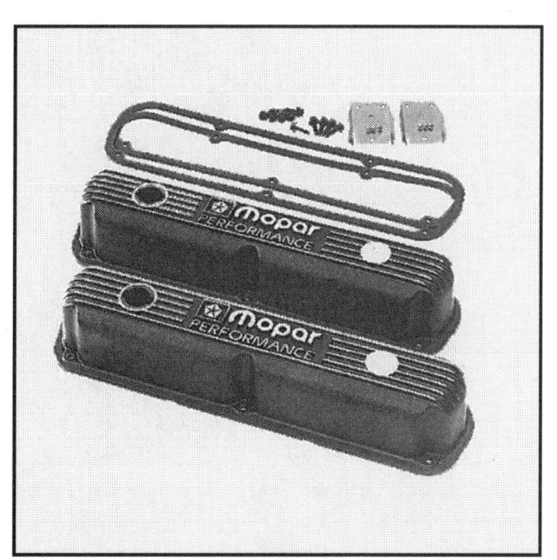

Both the Magnum 300 and the Magnum 380 engine assemblies cone outfitted with stylish and functional cast aluminum valve covers featuring black wrinkle finish and distinctive Mopar Magnum logo.

The package also includes a double roller timing chain and a MP electronic distributor. Distinctive cast aluminum valve covers are included and the oil pan is the center sump design similar to the Commando engines.

With the recommended 750 CFM Holley carburetor and headers, this package tested at over 380 horsepower at 5300 rpm and 410 lb-ft of torque at 4400 rpm. Over 350 lb-ft of torque is available from 2000 rpm to 5700 rpm. With only minor tweaking this is one of the few factory crate motors that can deliver 400+ horsepower out of the box.

Both of these engine can be easily installed using production motor mounts, starters and flywheels or flexplates.

Photos courtesy of Summit Racing

330 HP & 360 HP 350 Chevys

Summit Racing is the largest mail order speed equipment supplier in the country. Their engine shop is fully equipped to construct high performance engines for all applications. They offer a variety of long block and short block crate motors and a complete line of engine kits for home assembly.

Two basic 350 Chevy crate motors are offered They deliver either 330 or 360 horsepower from the same package with different camshafts. Both engines are equipped with Trick Flow Twisted Wedge aluminum cylinder heads sporting 2.02-inch intake valves and 1.60-inch exhaust valves. Available in either long block or short block form, these engines can be specified with either a driver's side or passenger side dip stick location, and the short blocks can also be ordered without a camshaft for those who wish who wish to make their own selection.

Each combination is built with a CNC-machined 4-bolt main 350 block, SummitEngine Shop high pressure oil pump, true roller timing chain, Clevite bearings and chrome valve covers.

The 360 horsepower model comes with TRW forged pistons and Stage 1 connecting rods with ARP bolts. The 330 horse engine uses high tech hypereutectic pistons with a D-shaped dish, moly rings and Stage 1 connecting rods.

A Solid foundation for each Summit crate engine includes the Engine Shop CNC machined cylinder block. These blocks are chemically stripped and pressure tested, fitted with all new hardware, your choice of OEM straight or Summit splayed main caps, align honed, decked and blueprinted for maximum performance.

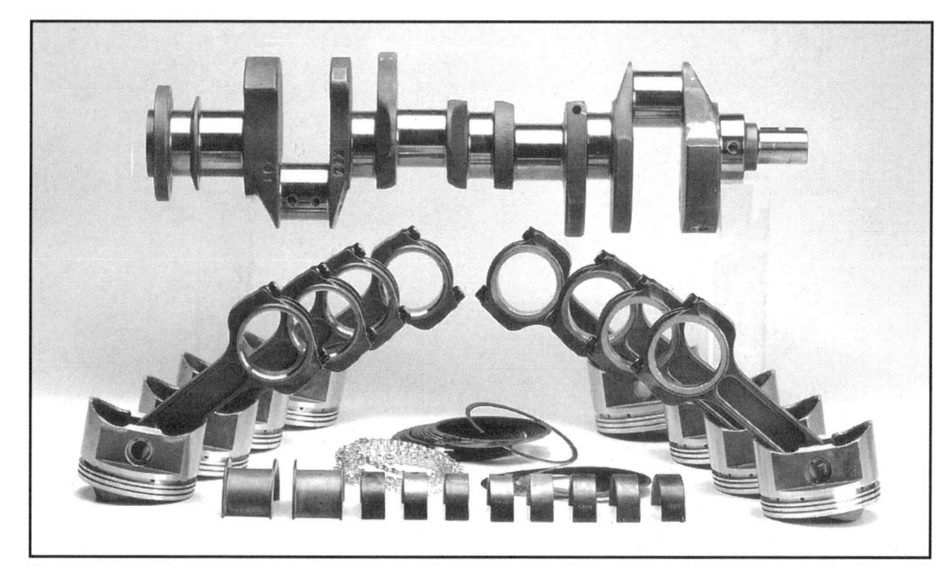

Summit Engine Shop balanced rotating assemblies are the heart of each Summit small block crate engine. Similar assemblies are also available separately for do-it-yourself engine builders. Separate assemblies include a Pro Line forged 350 or 400 crankshaft and connecting rods, Manley or TRW forged pistons, Speed Pro file-fit piston rings and Michigan 77 race engine bearings.

SUMMIT 360 HP CHEVY 350 V8 SPECS:

GENERAL:	
Part Number	**SES-3-48-99-310**
Displacement	350-cid, 5.7L
Bore & Stroke	4.00 x 3.48
Compression	9.5:1
Rated Power	363 @ 5500 rpm
Rated Torque	400 @ 3750 rpm
SHORT BLOCK:	
Block	4-bolt, CNC 350
Crankshaft	3.48" cast iron
Rods	Stage I, 5.7"
Pistons	KB hypereutectic
Rings	Moly
Bearings	Clevite
Balancer	not included
CAMSHAFT:	
Manufacturer	Summit
Type	Hydraulic
Valve Lift	.454/.454-inch
Adv. Duration	272°/.272°
CYLINDER HEAD:	
Type	Twisted Wedge
Material	Aluminum
Valve Material	Stainless Steel
Int. Valve Dia.	2.02
Exh. Valve Dia.	1.60
Head Bolts	ARP
Rocker Ratio	1.5

New Summit chrome valve covers are included with both long block assemblies and several accessory kits are available so you can complete the package. These include water pump kits with drive systems, induction kits with carburetor, intake air cleaner and hardware, fuel pump kits, ignition kits and harmonic balancer kits. Unless you can supply your own components, you'll need the items in these kits to get your crate motor up and running.

The 360 horsepower version has a more aggressive camshaft. It makes a good powerplant for hot street machines when coupled with 3.73 or numerically higher gears and a stick transmission or an automatic with a 3000 rpm torque converter.

Both engines deliver strong performance and both require the same combination of accessory components to install them. These include motor mounts, balancer, starter, flywheel or flexplate, bellhousing, intake system, exhaust system, ignition system, fuel lines, plugs and plug wires and all associated hardware.

450 HP 383 Chevy

The top of the line crate motor from Summit is the 450 horsepower 383 small block Chevy V8. This ground pounder includes options for a roller cam and a Victor Jr. intake manifold, and recommendations for a 750 CFM Holley double pumper, MSD-6A ignition, 1-3/4-inch headers, Flowmaster mufflers, 4.10 gears and a 3000 rpm stall torque converter.

The standard version features a Crane hydraulic camshaft with 286°/286° duration and .510/.512-inch valve lift. The optional Crane hydraulic roller cam offers 276°/284° duration and .488/.509 lift. If you select the option Victor jr. intake manifold it will come port matched to the aluminum cylinder heads.

The engine is equipped with a Summit roller timing chain, harmonic balancer and chrome valve covers

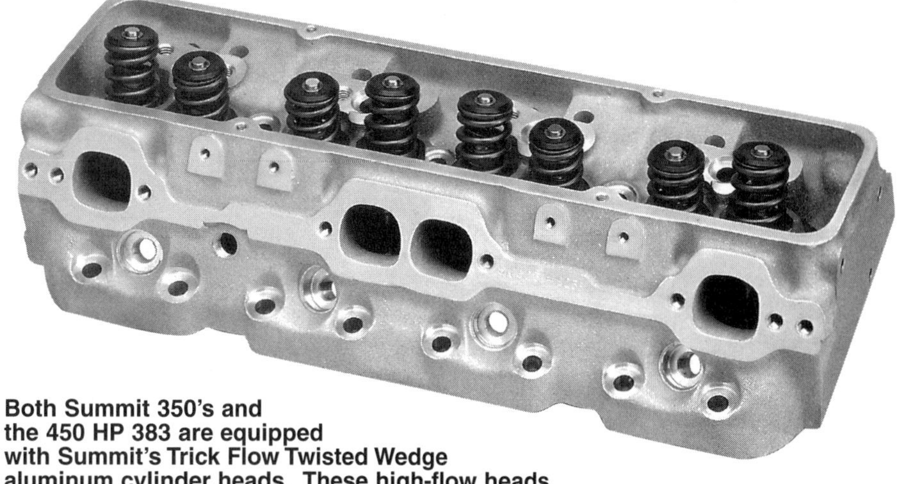

Both Summit 350's and the 450 HP 383 are equipped with Summit's Trick Flow Twisted Wedge aluminum cylinder heads. These high-flow heads feature 2.02-inch stainless stell intake valves and 1.600-inch stainless exhaust valves. They have 64cc combustion chambers and they accept most aftermarket headers and intakes. They also carry CARB EO number D-369 for use in California cars.

SUMMIT 330 HP CHEVY 350 V8

SPECS:

GENERAL:

Part Number	**SES-3-48-99-305**
Displacement	350-cid, 5.7L
Bore & Stroke	4.00 x 3.48
Compression	9.5:1
Rated Power	330 @ 5000 rpm
Rated Torque	390 @ 3500 rpm

SHORT BLOCK:

Block	4-bolt, CNC 350
Crankshaft	3.48" cast iron
Rods	Stage I, 5.7"
Pistons	KB hypereutectic
Rings	Moly
Bearings	Clevite
Balancer	not included

CAMSHAFT:

Manufacturer	Crane
Type	Hydraulic
Valve Lift	.440/.454-inch
Adv. Duration	210°/.216°
Lobe Centerline	114°

CYLINDER HEAD:

Type	Twisted Wedge
Material	Aluminum
Valve Material	Stainless Steel
Int. Valve Dia.	2.02
Exh. Valve Dia.	1.60
Head Bolts	ARP

with crankcase breathers. Like Summit's other engines, the 383 is precision assembled by the Summit Engine Shop. A CNC-machined 4-bolt main block is the startiong point. To this they ad a 3.75-inch stroke cast crankshaft, Stage II, 5.7-inch center-to-center length forged connecting rods and forged pistons to assemble the short block.

Accessory kits are also offered for the 383 engine. these include a water pump kit with either short or long style aluminum pump, aluminum drive pulleys, 180° thermostat, chrome housing and hardware. The ignition kit incorporates an MSD 8361 distributor with MSD cap and rotor, MSD-6A ignitionwith MSD Blaster-2 ignition coil, Moroso spark plug wires and wire separators and Accel spark plugs.

450 HP 383 cubic inch small block Chevy highlights the Summit Racing engine lineup. It is the ideal power plant for most hot street machines, street rods and lower level bracket racers.

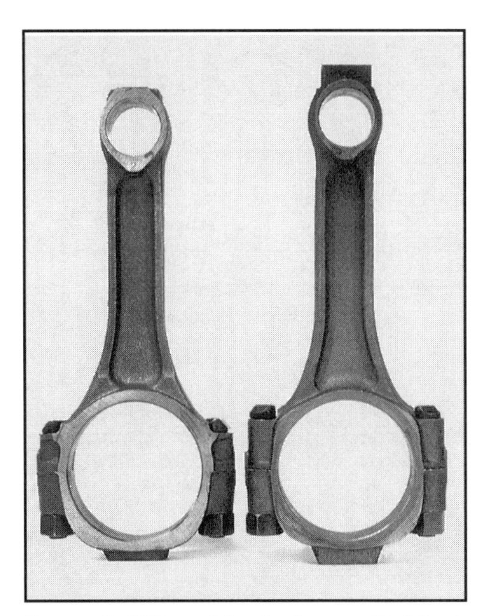

Summit 383 cubic inch small block Chevys are assembled with the longer 5.7-inch factory style connecting rod instead of the common .135-inch shorter 400 connecting rod (left). This puts less stress on the cylinder walls in a high performance application. These Summit Stage II rods are fully prepared and blueprinted.

SUMMIT 450 HP CHEVY 383 V8

SPECS:

GENERAL:

Part Number	**SES-3-51-99-455**
Displacement	383-cid, 6.3L
Bore & Stroke	4.030 x 3.75
Compression	9.5:1
Rated Power	450 @ 5750 rpm
Rated Torque	445 @ 4750 rpm

SHORT BLOCK:

Block	4-bolt, CNC 350
Crankshaft	3.75" cast iron
Rods	Stage II, 5.7"
Pistons	KB hypereutectic
Rings	Sealed Power
Bearings	Clevite
Balancer	Summit

CAMSHAFT:

Manufacturer	Crane
Type	Hydraulic
Valve Lift	.510/.512-inch
Adv. Duration	286°/.296°

CYLINDER HEAD:

Type	Twisted Wedge
Material	Aluminum
Valve Material	Stainless Steel
Int. Valve Dia.	2.02
Exh. Valve Dia.	1.60
Head Bolts	ARP
Rocker Ratio	1.5

Summit Engine Kits

Summiti also supplies complete engine kits for the d0-it-yourselfer whi wishes to assemble his own engine. You can get a 300 HP, 330 HP, or 360 HP 350 in long block or short block kit form, or you can purchase only the rotating assembly or bottom end kit. The 300 and 330 HP kits use cast aluminum pistons while the 360 HP kit uses KB hypereutectic pistons. All kits have a cast crankshaft and forged connecting rods.

For big block applications you can get kits for 396, 402, 427 and 454 engines with .010 or .020 -inch under cranks. A special 454 kit is also available with a steel LS-6 crankshaft and rods, a Summit steel balancer and the whole assembly is prebalanced. TRW forged pistons are use in all big block kits except the 402 which gets hypereutectic pistons.

Summit big block engine kits provide all the components you need to assemble your own high performance big block in a variety of displacements.

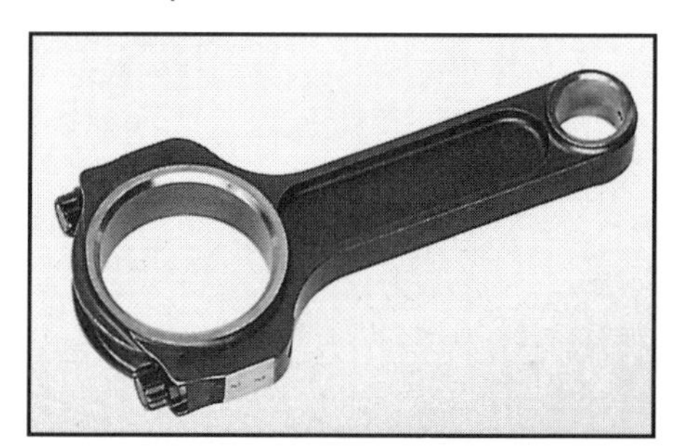

Small block engine kits from Summit Racing are available in short block and long block form to meet your every need. The compoents are pre-matched and will work perfectly if you provide careful, competent assembly.

Extra high performance applications may want to consider using the Summit Pro Line balanced rotating assemblies which feature TRW or Manley forged pistons, Manley steel rods and Pro Line forged crankshaft. These high performance assemblies are available for building high performance 355, 383 and 406 cubic inch small blocks and big blocks in 468, 496, 541, 572, 604 and 612 cubic inch versions.

SOURCE

SUMMIT RACING
P.O. Box 909
Akron, OH 44309-0909

24 Hr., 7 day order line
1-330-630-3030

Racing Head Service

Chevy 305/350 High Energy Engines

Racing Head Service has been one of the nation's premier high performance supplier's of engines and cylinder heads for decades. Their current lineup of high performance engines is one of the most complete arrays of performance engines available. At the core of the engine line are the High Energy 305/350 Chevy engines. Each of these engines features a bored cylinder block that has been torque plate honed to ensure good ring seal and the internal components are electronically balanced. All engines receive new World Products Street Replacement cast iron cylinder heads with steel exhaust seats to prevent seat recession with unleaded fuels. These heads have 1.94-inch intake valves and 1.5-inch exhaust valves.

The valve train includes all Competition Cams components and the engines are assembled using all new parts except the cylinder block, crankshaft and connecting rods. These engines are individually assembled and blueprinted using Silv-O-Lite cast pistons with nominal 8.8:1 compression ratio and Sealed Power moly rings. Each engine includes new chrome valve covers and timing cover, a new oil pan, new connecting rod bolts and a Melling oil pump. The accompanying dyno results illustrate a typical power curve generated by a RHS High Energy 350 small block equipped with a brand new Edelbrock aluminum Performer intake manifold, Edelbrock 600 CFM carburetor, HEI ignition and Hedman 1-5/8-inch headers.

305/350 cubic inch RHS High Energy engines are a staple of the high performance crate motor industry. These engines deliver good strong performance with solid reliability and stylish good looks.

HIGH ENERGY 305/350 SMALL BLOCK CHEVYS

Application	Pump*	Cam	RPM Range	Notes	Part Number
305/79 & back	yes	252 Hyd.	idle to 4500	AB	12305252H1
305/80-85	yes	252 Hyd.	idle to 4500	BI	12305252H2
305/86 & up	yes	252 Hyd.	idle to 4500	BDEFI	12305252H3
305/79 & back	yes	260 Hyd.	1400 to 5000	AB	12305260H1
305/80-85	yes	260 Hyd.	1400 to 5000	BI	12305260H2
305/86 & up	yes	260 Hyd.	1400 to 5000	BDEFI	12305260H3
305/79 & back	yes	268 Hyd.	1800 to 5500	AB	12305260H1
305/80-85	yes	268 Hyd.	1800 to 5500	ABI	12305260H2
305/86 & up	yes	268 Hyd.	1800 to 5500	ABDEFI	12305260H3
350/79 & back	yes	252 Hyd.	idle to 4500	AB	12305252H1
350/80-85	yes	252 Hyd.	idle to 4500	BI	12305252H2
350/86 & up	yes	252 Hyd.	idle to 4500	BDEFI	12305252H3
350/79 & back	yes	260 Hyd.	1400 to 5000	AB	12305260H1
350/80-85	yes	260 Hyd.	1400 to 5000	BI	12305260H2
350/86 & up	yes	260 Hyd.	1400 to 5000	BDEFI	12305260H3
350/79 & back	yes	268 Hyd.	1800 to 5500	AB	12305260H1
350/80-85	yes	268 Hyd.	1800 to 5500	ABI	12305260H2

* mechanical fuel pump boss

CRATE MOTOR
Performance
Tips & Tricks

HIGH ENERGY COMPUTER COMPATIBLE CHEVY V8s

	Pump*	Cam	RPM Range	Notes	Part Number
305/80-85	yes	Hyd.	idle to 5000	H	12305260AE1
305/1986	yes	Hyd.	idle to 5000	DFH	12308260AE1
305/1986	no	Hyd.	idle to 5000	DFH	12309260AE1
305/1986	yes	Hyd. Rlr.	idle to 5000	DFH	12310260AE1
305/1986	no	Hyd. Rlr.	idle to 5000	DFH	12311260AE1
305/1987 & up	yes	Hyd.	idle to 5000	DEFH	12312260AE1
305/1987 & up	no	Hyd. Rlr.	idle to 5000	DEFH	12313260AE1
350/80-85	yes	Hyd.	idle to 5000	H	12350260AE1
350/1986	yes	Hyd.	idle to 5000	DFH	12352260AE1
350/1986	no	Hyd.	idle to 5000	DFH	12353260AE1
350/1986	yes	Hyd. Rlr.	idle to 5000	DFH	12354260AE1
350/1986	no	Hyd. Rlr.	idle to 5000	DFH	12355260AE1
350/1987 & up	yes	Hyd.	idle to 5000	DEFH	12356260AE1
350/1987 & up	no	Hyd. Rlr.	idle to 5000	DEFH	12357260AE1

Notes:

A. Not computer compatible
B. World Products 305 SR Torquer Heads
C World Products 305 SR heads
D 1986 & later w/manual trans must have counterbalanced flywheel, extra charge
E Specify 1987 center intake bolt angle
F One-piece rear main seal
G Holes for Corvette useage must be redrilled & tapped
H Hypereutectic pistons
I '82 & later Z28 w/stock exhaust requires new GM oil pans
* mechanical fuel pump boss

High Energy engines are intended for moderate performance use in a variety of applications including street machines, street rods, towing and stock replacement in older cars. Computer compatible High Energy engines are also available for installation in late model cars requiring new or replacement engines. Some of these engines have different equipment based on the year and model required. Some use hypereutectic pistons and others require extra cost modifications to suit the specific application. The accompanying applications chart will guide you in choosing the best possible RHS engine.

High Energy engines are also available with various extra cost options. These include 1.52:1 ratio Magnum roller tip rocker arms, chrome oil pans, RHS Signature aluminum valve covers (short or tall), Edelbrock signature aluminum valve covers (short or tall), new 350 Chevy automatic transmission flexplate, or 153 tooth 400 flexplate, Hays 30 pound standard transmission flywheel for Chevy 350 with pilot bearing installed and new bolts, Hays 30 pound 400 manual transmission flywheel with pilot bearing installed and new bolts or a Pioneer OEM style manual transmission flywheel with pilot bushing installed.

Chevy 383/400 High Energy Engines

High Energy engines are also available in 383 and 400 cubic inch versions. They include all the same basic features as the 305/350 series engines, but they are ideal for heavier vehicles or applications where the additional performance available from extra displacement is required. These engines are built with either 5.565-inch connecting rods or 5.7-inch connecting rods depending on the application. Like the 305/350 engines, they are available with 252, 260 and 268 High Energy Cams which give them a performance range from off-idle to as high as 5400 rpm.

Chevy 350/383/400 Magnum Engines

RHS Magnum engines are the next level of performance for those who seek a very strong performance package that is still streetable. they are built with stronger internal components to take more abuse and the 383/400 versions are built in both long and short rod versions to tailor the torque and horsepower curves to individual requirements.

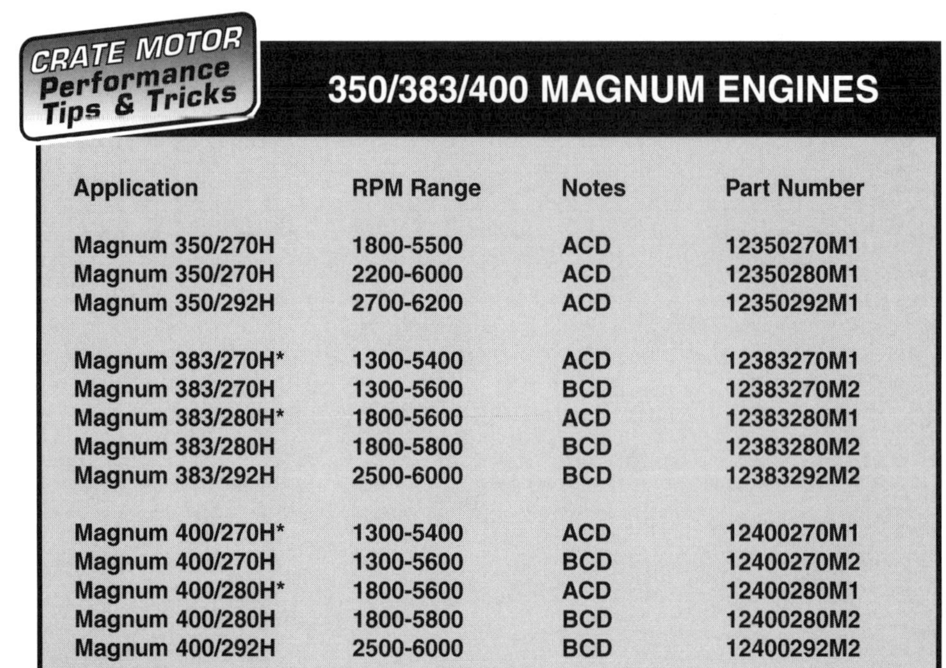

CRATE MOTOR Performance Tips & Tricks

350/383/400 MAGNUM ENGINES

Application	RPM Range	Notes	Part Number
Magnum 350/270H	1800-5500	ACD	12350270M1
Magnum 350/270H	2200-6000	ACD	12350280M1
Magnum 350/292H	2700-6200	ACD	12350292M1
Magnum 383/270H*	1300-5400	ACD	12383270M1
Magnum 383/270H	1300-5600	BCD	12383270M2
Magnum 383/280H*	1800-5600	ACD	12383280M1
Magnum 383/280H	1800-5800	BCD	12383280M2
Magnum 383/292H	2500-6000	BCD	12383292M2
Magnum 400/270H*	1300-5400	ACD	12400270M1
Magnum 400/270H	1300-5600	BCD	12400270M2
Magnum 400/280H*	1800-5600	ACD	12400280M1
Magnum 400/280H	1800-5800	BCD	12400280M2
Magnum 400/292H	2500-6000	BCD	12400292M2

Note: * 5.565-inch rods

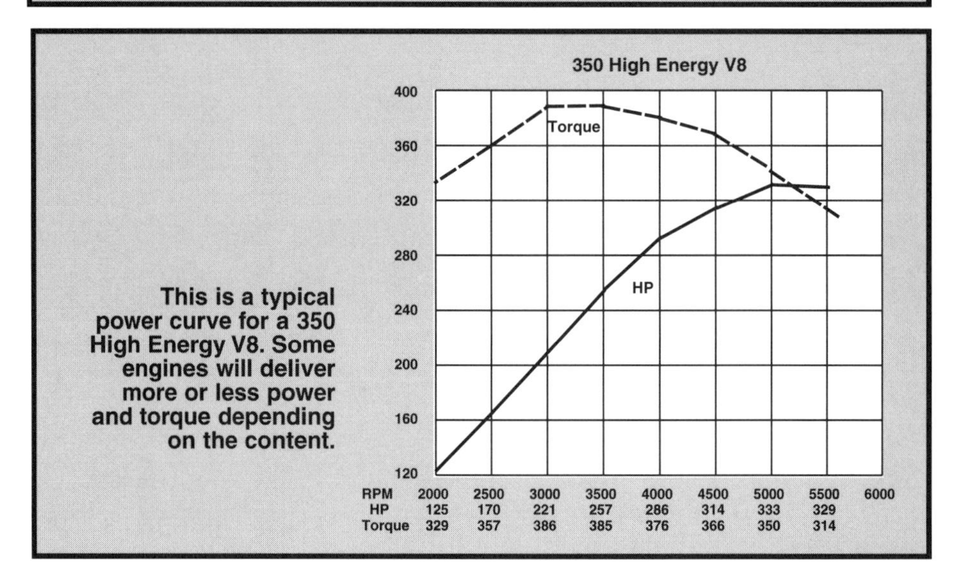

This is a typical power curve for a 350 High Energy V8. Some engines will deliver more or less power and torque depending on the content.

These engines use World Products Street Replacement cast iron cylinder heads with bronze valve guides and 2.02/1.600-inch stainless steel valves, bowl porting and intake port matching. The valve train is supplied by Competition Cams and the pistons are either forged or hypereutectic

CRATE MOTOR Performance Tips & Tricks

350/383 MAGNUM SPECIAL ENGINES

Application	Comp Ratio	Notes	Part Number
Magnum 350/280H	9:1	AC	12350280S
Magnum 350/292H	9:1	AC	12350292S
Magnum 350/292H	10:1	AC	12350292SH
Magnum 350/305H	10:1	AC	12350305SH
Magnum 383/280H	9:1	BCD	12383280S
Magnum 383/292H	9:1	BCD	12383292S
Magnum 383/292H	10:1	BCD	12383292SH
Magnum 383/305H	10:1	BCD	12383305SH

RHS also offers their Magnum Special engines incorporating the Edelbrock Performer RPM power package. This includes Edelbrock aluminum heads, cam and lifters, Performer intake, Melling oil pump and timing set and Comp Cams roller tip rockers. The Performer Magnum Special can be applied to 350 and 454 Chevys, Ford 302 and 351W engines and Pontiac 400s and 455s.

Magnum Special

Magnum Special engines are extra high performance versions equipped with Dart Sportsman cylinder heads, different cam profiles and higher compression ratios. These combinations are designed to deliver maximum power for street and occasional strip use. Magnum Special engines feature bowl-ported and port-matched cylinder heads, ARP head bolts with hardened washers, Cloyes True Roller timing sets, forged or hypereutectic pistons, ARP main studs and rod bolts and all the same options included in the standard magnum packages. Some applications may also want to check into the Chevy S/B Hy-Comp engines which are specially configured Magnum Special engines that take maximum advantage of combustion chamber shape and piston deck to produce maximum power on standard pump gas.

Chevy 350/383/400

depending on the application. Moly piston rings are used and each engine is assembled with high strength ARP main studs and rod bolts.

All engines have chrome valve covers, timing cover and a new oil pan, Melling oil pump, double row timing chain and Comp Cams 1.52 ratio Magnum roller rockers. The 383s have a new cast stroker crank. Additionally, these engines have most of the same optional equipment available as the 305/350 series High Energy engines, plus the availability of World Products Sportsman Heads with 2.02/1.60 valves, guide plates, hardened pushrods and a choice of 64cc or 72cc chamber volume and angle or straight plug orientation. Or you can select Edelbrock Performer RPM cylinder heads with CNC porting. See the accompanying chart for magnum Engine applications. RHS Magnum engines are very strong performers when installed in typical street machines, street rods and some kit cars. Some enthusiasts have even installed them in marine applications with great success

Street rodders will be interested in the Chevy 350/383 Street Rod Special engine which features RHS prepared Corvette aluminum cylinder heads and hyperentectic pistons. Intake, carb, ignition and headers are optional.

Chevy 350/383

Upon customer's request, an engine can be configured with Accel EFI electronic fuel injection system. Additional components are required to run the electroncis. RHS can supply everything you need.

High Performance Specials

These are the top of the line small blocks from RHS. They feature ported Dart Sportsman or Edelbrock Victor aluminum cylinder heads, aluminum needle bearing roller rocker arms, new steel crankshafts and forged pistons. These are very high performance engines and you can specify such items as a Pete Jackson gear drive, Stainless Steel Comp rocker arms, a Fluidampr harmonic balancer, BRC or Callies 4340 alloy crank, Eagle connecting rods or Oliver billet connecting rods and a Centerforce flywheel. Optional intakes include a port matched Victor Jr., a CNC ported Victor Jr. or Brodix HV1 intake. These engines have 10:1 to 12.5:1 compression ratios and they require racing fuel. Roller or flat tappet camshafts are used in all of these high performance engine packages.

Obviously this type of engine is a custom package and RHS is capable of specing a complete custom package to meet your needs. If you don't see what you want, just ask. They can probably do it. In these cases the engines will be considerably more expensive due to the cost of specialty parts and the cost of making these various components work in harmony with the rest of the package.

Street Stock Chevy 350 Engines

Custom built 11:1-plus compression ratio small blocks designed for Street Stock, Hobby Stock, late Model, Modifieds, Modified Affordable, IMCA and most local tracks. Available with either mechanical or hydraulic camshaft, these 400 plus horsepower engines are built tough to take the punishment of racing applications.

They feature ARP rod bolts and main sets, oval track wet sump oil pan and pickup screen, Comp Cams TL series cam and 4-bolt prepped cylinder block. The cylinder heads are selected from good 291, 461 or 462 head castings that are surfaced and configured to yield 11:1 compression ratio. Manley severe duty 2.02 and 1.60-inch diameter stainless steel valves are used with steel retainers and Comp Cams 10 degree locks.

Each engine is electronically bal-

A Supercharged small block 350 called the 350 BDS engine is based on the RHS High Performance Special Engine.

anced and the cast crankshaft is indexed. A new GM harmonic balancer and roller timing set are included and special oval track valve covers with breathers are incorporated for the racing application.

Combinations can be adjusted to suit your local track rules. when ordering, just inform the salesman of any specific bore and stroke rules or any other engine restrictions that must be met for your track and racing classification. Keep in mind that racing fuel must be used with these engines and that increased maintenance is required to keep them running at top performance levels.

Chevy Big Blocks

RHS big blocks are available in four popular sizes including 396, 402, 427 and 427 cubic inch models. The High Energy series engines are designed for brute low end torque and mid-range power and they are generally compatible with standard pump gas.

For high performance street applications they offer the Magnum Special series designed for big power gains in the mid to upper rpm range. the combinations offer a choice of camshaft and compression ratio to assure the streetability you need for your applications.

All RHS big block combinations use oval port cylinder heads to maximize torque in the street rpm range. Magnum Special versions use new World Products Merlin oval port heads for maximum breathing with the Comp Cams Magnum camshaft grinds.

If you select a High Energy Series performance big block engine it will include bronze valve guides and hardened exhaust seats for use with standard unleaded pump gas. The cylinder heads are bowl-ported and fitted with 2.065/1.72-inch stainless steel valves and a complete Comp Cams valve train. Silv-O-Lite cast pistons are teamed with moly piston rings for good street performance. Each High energy engine also includes brass freeze plugs, chrome valve covers, chrome timing cover, Melling oil pump and double row timing chain and new high performance connecting rod bolts.

Big block enthusiasts can also opt

RHS big block Chevy engines are offered in 396, 402, 427 and 454 cubic inch sizes with a variety of options for camshafts and induction systems. Other options include EFI systems, Brodix aluminum heads, roller rockers, 4340 steel cranks, forged pistons,Eagle Specialty rods, Fluidamprs and Moroso oiling systems.

CRATE MOTOR Performance Tips & Tricks

BIG BLOCK APPLICATIONS

Application	RPM Range	Notes	Part Number
High Energy 260H 396/402	idle-4500		11396260H
High Energy 260H 427	idle-4500		11427260H
High Energy 260H 454	idle-4500	A	11454260H
High Energy 268H 396/402	idle-5000		11396268H
High Energy 268H 427	idle-5000		11427268H
High Energy 268H 454	idle-4800	A	11454268H
High Energy 270H 396/402	idle-5500		11396270H
High Energy 270H 427	idle-5200		11427270H
High Energy 270H 454	idle-5000		11454270H
Magnum Special 270H 427	1400-5500	D	11427270S
Magnum Special 270H 454*	1300-5300	ACD	11454270S
Magnum Special 280H 427	2000-6000	D	11427280S
Magnum Special 280H 454*	1800-5800	ACD	11454280S
Magnum Special 280H 454**	1800-5900	ACD	11454280SH
Magnum Special 292H 427	2800-6300	D	11427292S
Magnum Special 292H 454*	2400-6000	ACD	11454292S
Magnum Special 292H 427**	2400-6200	ACD	11454292SH
Magnum Special 305H 427	3200-6800	D	11427305S
Magnum Special 305H 454**	3000-6700	ACD	11454305SH
Magnum Special 268H 427	3000-6500	D	11427268S
Magnum Special 268H 454*	2200-6000	ACD	11454268S
Magnum Special 268H 454**	2200-6000	ACD	11454268SH

A Externally balanced engine includes auto trans flexplate
C Compression ratios are approximate
D Includes Competition Cams Magnum roller tipped rockers

* 9:1 CR ** 10:1 CR

for the big block 454 High Performance Series. These engines are all out performance street/strip powerplants built with the high quality components necessary to produce the power and reliability this application requires. Big Block HPS combination number 1 features the Magnum Special build plus ported Merlin heads, 10:1 TRW forged pistons, RHS magnum rods with ARP Wave-lok bolts, Comp Cams 292 camshaft, Cloyes true roller timing chain, Comp Cams Magnum roller tipped rocker arms with hardened pushrods and a Milodon or Moroso street/strip oil pan that fits most popular chassis.

A second HPS combination includes all of the above plus a Competition Cams 294S solid lifter cam and RHS aluminum 1.7 ratio roller rockers. The 454 HPS #1 combination is PN 11454HPS1 while 454 HPS # 2 combination is PN 11454HPS2..A lengthy list of options is also offered for these engines. Check with your salesman for exact components available.

Chevy Big Block 572 HPS

If you're in the market for a thumper "mountain motor" setup, the 572 HPS is your answer. This engine produces 750 lbs.-ft of torque and RHS claims it can almost move mountains. Based on the HPS High Performance Series, the 572 is built from all new aftermarket parts. The camshaft and compression ratio can be specified for different applications.

The 572 HPS includes an all new GM Bowtie block, BRC 4340 non-twist forged steel crankshaft, Oliver Parabolic Beam billet steel rods, JE custom forged pistons, Sealed Power plasma moly rings, Dart Merlin rectangular port cylinder heads, Comp Cams roller camshaft and valve train, Manley severe duty valves, Moroso stroker oil pan, Fluidampr harmonic balancer, SFI approved flexplate and RHS signature valve covers. To order this engine from RHS ask for the 572 HPS under PN RHS11572BRK.

Chevy Big Block Blower Engine

The blower engine is designed for all-out Pro Street and Street Rod applications. It displaces 454 cubic inches and includes a new GM steel crank, new Eagle Specialty 4340 rods, JE forged blower pistons, Comp Cams roller camshaft and valve train, Pro Magnum roller rockers, aluminum timing chain cover, O-ringed block with copper head gaskets, Manley severe duty stainless steel valves,Merlin rectangular port cylinder heads, bowl porting and full port matching, Moroso 7 quart oil pan, RHS signature series valve covers, BDS polished blower with calibrated dual Holley carbs, fuel lines and scoop with filters, Fluidampr harmonic balancer with ARP bolt, MSD magnetic trigger distributor and MSD 6-BTM control box, Blaster -2 ignition coil, bronze distributor gear and plug wires. Ask for PN RHS11454300BDS.

Chevy 350 "Legal Eagle" Engine

RHS also offers a basic environmentally friendly Chevy 350 engine package using all emission legal parts. This engine propelled their '87 Z28 Camaro project car to 12.50 elapsed times at 108 mph without the aid of nitrous oxide injection or any boost from a supercharger. This basic replacement 350 features KB hypereutectic pistons, Total Seal Gapless piston rings, RHS prepared Edelbrock or Air Flow Research aluminum cylinder heads, Special Grind hydraulic roller camshaft and a late model cylinder block with one-piece rear main seal.

You simply transfer your existing induction and exhaust system and all accessories to the new engine and you've got a plenty hot street engine that won't pollute the air. This engine showed 0.0% carbon monoxide and 34 parts per million hydrocarbons on a recent emissions test. To enjoy the power and performance and retain clean emissions ask for PN RHS12350MLE.

Ford 302/351 Engines

Ford 302 and 351 engines are also offered in the same High Energy, Magnum, Magnum Special and High Performance Series HPS configuration as Chevy engines. High Energy versions feature torque plate honed cylinder blocks, balanced assemblies, cast pistons, Comp Cams valve train, bowl ported heads, Sealed Power moly rings, Melling oil pump, harmonic balancers and flexplate, chrome valve covers and brass freeze plugs.

Magnum versions add TRW forged pistons, intake port matching, APR main cap studs and rod bolts, and Comp Cams Magnum roller rockers. Ford Motorsport GT-40 cast iron or aluminum heads are optional and you can also specify and Centerforce 30-pound steel flywheel, Cloyes True Roller timing chain, Fluidampr balancer and a rear sump oil pan.

If you select a Magnum Special engine you gain RHS prepared World Products Windsor Jr. (302) or Windsor (351W) cylinder heads, Comp Cams Magnum roller rockers, guide plates and ARP rocker studs and hardened pushrods. Options with this package include a solid lifter camshaft, World Products aluminum heads, Edelbrock Performer RPM aluminum heads with 2.02/1.6 valves, Harland Sharp roller rockers with Comp Cams Hi-Tech pushrods, RHS aluminum roller rockers, Comp Cams Hi Tech stainless rockers and a Pete Jackson gear drive.

High Performance Special HPS engines feature custom lightweight forged pistons (10.5:1 CR), Comp Cams 290-B6 flat tappet camshaft and valve train components, RHS prepped Windsor Magnum heads, index ground crankshaft, RHS Super Stock rods with ARP bolts and RHS aluminum roller rocker arms.

Or you can select HPS version 2 which offers 11.5:1 CR forged pistons, Comp Cams 296-BR6 roller camshaft RHS prepared Windsor cylinder heads, Fluidampr harmonic balancer and RHS aluminum roller rockers with a stud girdle. These HPS engines offer an extended rpm range for use with 4-speeds, high-stall converters and low ratio rearends.

5.0L EFI Magnum & Magnum Plus Engines

Designed as more powerful replacement engines for GT and LX Mustangs, Thunderbirds and trucks, these engines feature proven mix of high performance aftermarket parts. Magnum versions incorporate a RHS blueprinted block, GT-40 iron heads with Magnum porting,screw-in studs, guide plates and roller rockers, Comp Cams hydraulic roller valve train, TRW forged pistons, RHS magnum rods with ARP bolts and a Fox-body compatible rear sump oil pan and pickup. They deliver 320 horsepower and 359 lbs-ft. of torque for a smooth powerful daily driver.

The 414 stroker engine delivers all the performance you could ask for in a streetable package. Horsepower exceeds 400 with up to 470 lbs-ft. of torque available. This engine offers big block power in a compact small block package.

Magnum and Magnum Plus 302/351 EFI engines offer up to 350 horsepower and 370 lbs-ft of torque in a smooth, drivable package for late model Mustangs, T-Birds and pickups.

Magnum Plus versions also add Edelbrock Performer aluminum cylinder heads, Comp Cams, 1.7 ratio aluminum roller rockers and Comp Cams special EFI grind camshaft. This combination delivers 350 horsepower and 370 lbs-ft of torque. It can also be configured for an add on centrifugal supercharger if you desire.

414W Stroker Engine

The 414 cubic inch Windsor stroker engine offers 400-plus horsepower and 470 lbs-ft. of torque with good drivability and a decent idle. The 9:1 CR version incorporates RHS prepped Magnum Dart Windsor cylinder heads, RHS 351W block, index ground crankshaft, forged pistons, Eagle Specialties 4340 steel connecting rods, custom grind Comp Cams hydraulic camshaft, Pro magnum roller rockers ARP main cap studs and ARP head bolts with hardened washers.

Options include a Comp Cams solid lifter cam up to .570-lift, World Products aluminum ,Fluidampr harmonic balancer, rear sump pan and pickup,Comp Cams Hi-Tech stainless steel rocker arms and pushrods and a Cloyes True Roller timing chain.

If you select the High Performance Special, HPS version you also get Fel Pro's Wire-Loc cylinder head gaskets, Manley severe duty stainless steel valves, A new Comp Cams roller cam and valve train and 9:1 or 10.5:1 forged pistons.This is a very hot package for any Ford application.

Racing Head Service engines have a good reputation. They have been used in countless magazine project cars and in thousands of high performance street machines, street rods, race cars, tow vehicles and marine applications.

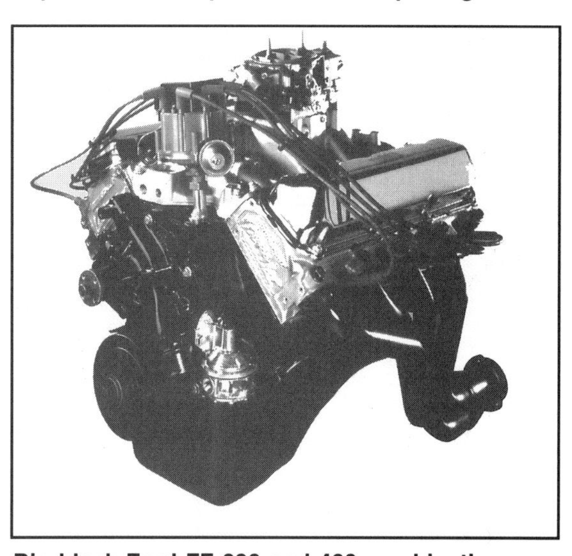

Big block Ford FE 390 and 460 combinations are also offered along with extra high performance 514 cubic inch stroker big blocks for the uiltimate in Ford performance.

SOURCE

RACING HEAD SERVICE
3416 Democrat Road
Memphis, TN 38118

Order line (800) 333-6182
Tech Line (901) 794-2830
Fax (901) 794-2838

AutoCenter is a large mail-order engine supplier centrally located in Dallas, Texas. They offer a variety of stock and high performance Chevy small blocks for use as replacement engines and for high performance applications. Depending on the application, some engines come more fully dressed than others. Some will include a water pump, harmonic balancer and intake manifold while others will not have these components because they are specified as replacement engines where many of these components will be swapped over from the previous engine.

Among their more basic engine packages is the 265 horsepower 350 small block shown on the right. This replacement engine delivers a solid 350 lbs-ft. of torque and is specified as an emission legal replacement engine for light duty 1973-85 applications. It is built from all new parts and includes a 4-bolt main cylinder block, 8.5:1 compression ratio cast pistons, stock oil pump, oil pan, valve covers, dual dip stick cylinder block, 76cc chamber cylinder heads with 1.95-inch diameter intake valves and 1.50-inch exhaust valves. If used in a light duty application, this engine comes with a nationwide 36 month, 50,000 mile warranty. While not necessarily a firebreather, this engine can probably be made to deliver close to 300 horsepower if you add a Edelbrock Performer intake manifold and a Q-Jet carb and headers.

265 horsepower basic replacement 350 engine from the AutoCenter has plenty of hop-up potential. With 4-bolt mains and medium size 1.94/1.50-inch valves it should breath well. A few bolt-ons and maybe a cam change and you're looking at 300 HP.

270 HP 350 Chevy

This is a all new replacement engine specified for 1987-92 Chevy truck applications. Two versions are available—one for light duty applications (NA1 option) and one for medium duty applications (NA4 option). Both engines deliver 270 horsepower at 4500 rpm and 350 lbs-ft. of torque at 3500 rpm when run with 1-5/8-inch diameter tubular headers.

These engines feature a beefy lower end with nodular iron crankshaft, one piece rear main seal, 8.5:1 compression ratio cast pistons, stock oil pump, oil pan, valve covers and leak resistant neoprene seals. The iron cylinder heads have 76cc combustion chambers with 1.94/1.50-inch diameter valves. Either engine is very suitable for replacement use or as a foundation for a mild performance street machine or street rod application. By adding appropriate bolt-on speed equipment both engines could easily approach 300 horsepower.

Both engines include a harmonic balancer and you can add a aluminum intake, high energy ignition source plus the carburetor and headers of your choice to design a pretty healthy street engine on a budget. Best of all, these engines also carry a 36 month, 50,000 mile warranty.

The 270 horsepower truck replacement engine offers all the right pieces for a good performng street engines in mildly modified street machines and street rods. With minor bolt-ons it can deliver 300 horsepower and years of trouble free service.

320 HP 350 Chevy

If you're ready to step up to a more power-specific engine, the AutoCenter 320 horsepower 350 with aluminum cylinder heads might be your best choice. This engine is specified for pre-1985 light duty applications, but it includes Corvette aluminum cylinder heads with 1.94/1.5-inch diameter valves and a 9.5:1 compression ratio.

The 4-bolt main blocks holds a standard nodular iron crankshaft with the earlier style 2-piece real main seal. The block features dual dip stick locations, and it comes with oil pump , oil pan and valve covers. Like the other small blocks in AutoCenter's lineup, this medium performance V8 can be easily hopped up with existing aftermarket performance hardware. Even with the higher output of this engine, it is still backed by a great 12 months parts warranty.

Because this engine is equipped with Corvette aluminum heads, it has a higher compression ratio than the iron head engines. This is possible because the aluminum radiates heat more easily. It permits more than a full point increase in compression ratio which does great things for power and it allows the use of a more aggressive camshaft without sacrificing low end power.

Factory style center-bolt valve covers are used on this engine, but you can easily replace them with aftermarket covers that complement the aluminum cylinder heads and the higher performance intent of your particular application.

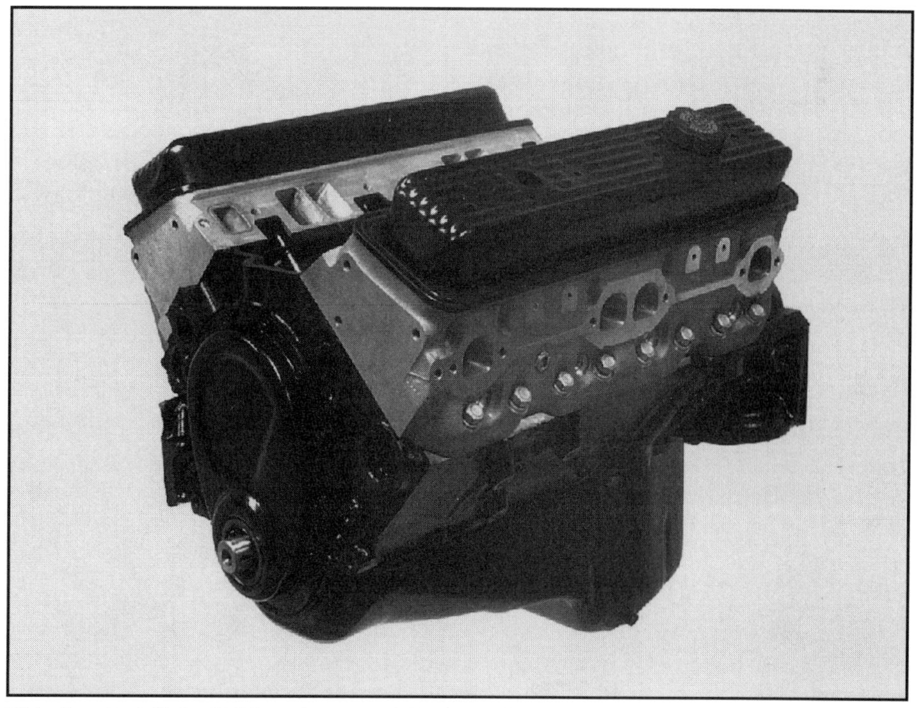

This is a hot little 320 horsepower 350 that will put some life into any street application. It makes power to 5000 rpm and beyond and delivers 370 lbs-ft, of torque. With 4-bolt mains and aluminum heads it makes a great basic performance engine.

375 HP, 350 Chevy Twisted Wedge Head

The "Twisted HO" is Chevy's ZZ3 HO 350 engine modified by AutoCenter with Trick Flow Specialties "Twisted Wedge" aluminum big valve cylinder heads. These heads have 2.02/1.60-inch canted valves with high performance valve springs.

The "Twisted HO" delivers 375 horsepower at 5700 rpm and over 400 lbs-ft. of torque at 4200 rpm. This power level remains reliable with the use of a 4-bolt main cylinder block, forged steel crankshaft, forged "Pink" coded connecting rods and 9.8:1 compression ratio hypereutectic pistons.

The engine comes equipped with a dual plane intake manifold, harmonic balancer, long style water pump and HEI electronic distributor. The camshaft is a dual pattern hydraulic roller with .474/.510 valve lift. It produces 16 inches of idle vacuum which will support most fuel injection systems with ease if you choose to add EFI.

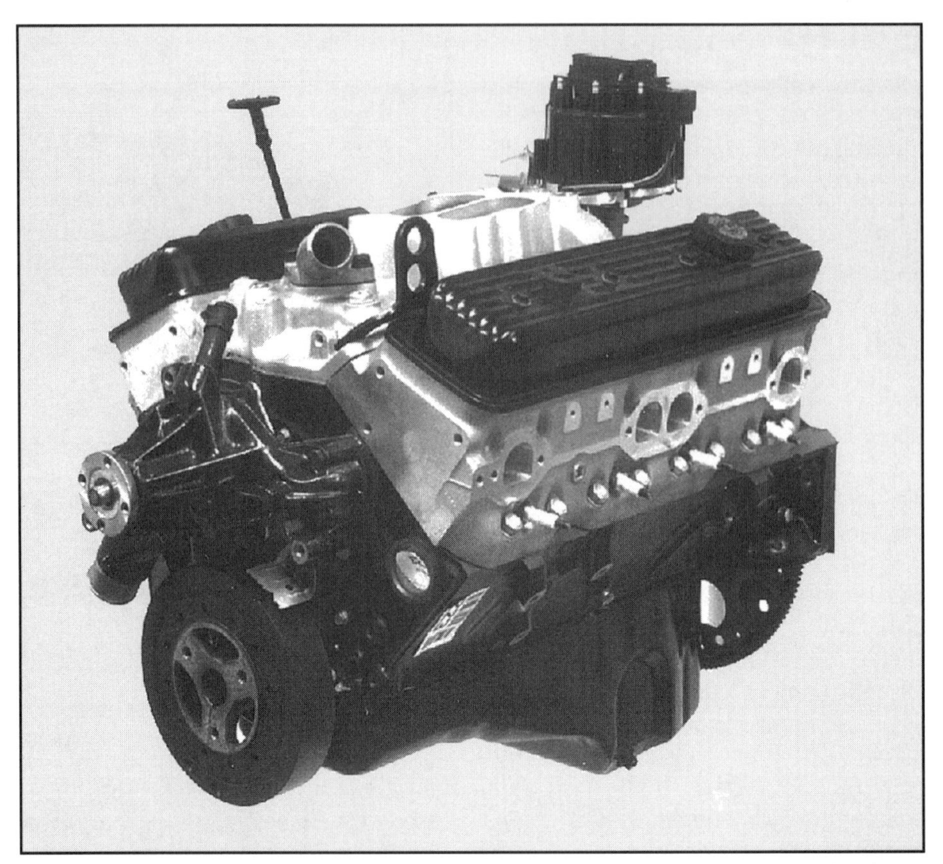

Here's a true hot rodded engine for hot street machine and street applications. This 375 horsepower 350 is equipped with aluminum Twisted Wedge cylinder heads, dual pattern camshaft, dual plane intake manifold, 4- bolt main block and heavy duty "Pink" connecting rods. It is AutoCenter's modified version of the GM ZZ3 HO 350 engine.

Carbureted GEN II LT1 370 HP 350

AutoCenter's hot new GEN II LT1 Chevrolet small blocks are specially modified versions designed to accomodate those enthusiasts who wish to run a carburetor, yet still benefit from the high technology incorporated in the new GEN II small block V8s.

Two versions are available—a 370 horsepower model that delivers 390 lbs-ft. of torque at 4000 rpm, and a 340 horsepower model that makes 370 lbs-ft. of torque at 4200 rpm. Each engine comes complete with an aluminum intake manifold, roller cam distributor gear and harmonic balancer.

They also incorporate GEN II technology with aluminum cylinder heads and front mounted distributor. The aluminum intake accepts all standard flange Holley carburetors and the mounting flange is also drilled to accept a Quadrajet four barrel or an Edelbrlck or Carter carburetor. The LT1 aluminum cylinder heads will accept all popular headers

If you want the added technology of later model GEN II engines, the AutoCenter offers two versions of the 350 engine with either 340 horsepower or 370 horsepower. Both engines produce over 16-inches of idle vacuum, making them entirely suitable for any street application.

AutoCenter's big 502 big block Chevy offers high torque and exceptional performance in an over-the-counter or mail-order package. Aluminum oval port heads and a mild hydraulic roller cam team up to deliver 550 horsepower and 570 lbs.-ft of torque.

550 HP, 502 Big Block Chevy

For many enthusiasts, big block power is still the only way to go. AutoCenter addresses their needs with a pair of big guns displacing 502 cubic inches. These engines differ slightly from the standard Chevrolet crate motors because they are built with a mixture of high performance aftermarket parts applied to a 502 short block.

Version one has the 4-bolt main short block with 1053 alloy steel crankshaft, 4340 connecting rods and forged pistons. This engine has Mark V oiling with integral oil cooler connections. The cylinder heads are the highly regarded World Products Merlin oval port castings which make plenty of power. It also includes a cast aluminum 502 valve covers, harmonic balancer, 14-inch flexplate and a 6 quart truck oil pan.. With a mild hydraulic roller camshaft this engines produces 570 lbs-ft. of torque at 4500 rpm and 550 horsepower at 5500 rpm. More than 500 lbs-ft. of torque is available at 2500 rpm.

The second version of the 502 big block is very similar except that it is equipped with a lightweight set of oval port aluminum big block cylinder heads. The lower end is the same: the rods have heavy duty 7/16-inch bolts and the forged pistons offer 9.25:1 compression ratio.

Power and torque are rated the same as verison one, but street machine and street rod enthusiasts are sure to prefer the lighter weight of the aluminum cylinder head version. Both engines have very strong internals that can take plenty of punishment.

Depending on the application you can outfit either of these engines with a variety of aftermarket speed components that will deliver top dog performance. Numerous oval port intake manifolds are available to fit the cylindert heads including EFI setups. Most popular header sizes will bolt right up and since the engines come with flexplates you are ready to attach them to an automatic trans with no particular problems. You'll need a water pump and pulleys, motor mounts and a dip stick. There is no boss for mounting a mechanical fuel pump so you will have to use an electric pump. With the recent introduction of GEN VI cylinder blocks, these engines may be assemble from GEN VI compoents in the future. That would suggest the future availability of a mechanical fuel pump if you really want one.

American Speed builds a variety of strong Chevy engines from 275 horsepower budget "Saturday Night Specials" to 750 horsepower 406 nitrous killer motors. American Speed performs all machine work and assembly in their own facility. All American Speed engines are balanced, blueprinted and dyno tested with each customer receiving the dyno test results for their engine. Each engine is dyno tuned to ensure power levels, reliability and engine break-in under a controlled environment. Engines are checked on the dyno, then disassembled and checked again prior to shipping to the customer. More than 40 different engine packages are available. These range from near stock Saturday Night Special 350s to fire-breathing nitrous small blocks and 550 horsepower 502 big block Chevys.

Saturday Night Special

These engines are built with all new parts, including 4-bolt main blocks, cast crankshafts, cast pistons and 8.5:1 compression ratio. You can buy them in either long block or fully assembled versions. Long block versions have a GM camshaft and no intake manifold while fully assembled versions come with a Crane Cams

American Speed's "Saturday Night Special" is a 275 HP entry level powerplant built with all new parts. It comes with a hot Crane cam and a Edelbrock Performer intake manifold.

284°/.450-lift cam, a new Edelbrock Performer intake manifold, harmonic damper, and chrome valve covers. A 3 year, 50,000 mile warranty covers all "Saturday Night Special" engines. This engine is rated at 275 horsepower with the components listed here. It is a good starting point for any high perfmance street machine project, or for towing applications and even moderate marine use. Many enthusiasts don't realize that these budget-built crate motors will deliver exceptional performance and service if they are well maintained.

400 HP 383 Chevy Enforcer

This is a high performance street engine for cars and trucks. It has 9:1 compression and it makes 16 to 17-inches of vacuum at idle with a slight lope. It will perform well on pump gas. A new cast crank is used along with ported Stage II World Products S/R Torquer heads. You can also get Trick Flow Twisted Wedge heads (shown on opposite page at the top). The pistons are forged TRW's with reconditioned rods, ARP bolts and a 4-bolt main block. The camshaft is a .480-inch lift Crane hydraulic unit and the engine comes with a Edelbrock Performer manifold, balancer, 6 quart pan and chrome valve covers. A 375 horsepower version is also available and the 400 HP 355 cubic inch version is also popular.

445 HP 383 Chevy Torquer

American Speed also offers a 445 horsepower 383 that has a racy idle, but still makes 15 inches of manifold vacuum at 1000 rpm. It uses lightweight TRW forged pistons, a /490-lift Competition Cams camshaft, an Edelbrock Performer RPM intake, Crane roller tipped rockers and 2.02/1.60 stainless steel valves in Stage II ported AFR aluminum cylinder heads. This engine is built around a 4-bolt main cylinder block with a nodular iron crankshaft, GM balancer and flexplate. You can get this engine either carbureted or with Edelbrock fuel injection if you wish.

445 HP 383 is available both carbureted and fuel injected with Edelbrock Fuel Injection. These engines feature forged TRW pistons, high lift Competition Cams camshaft and full dyno testing.

450 HP 406 Chevy Torquer

A 406 cubic inch carbureted engine is another popular choice in the American Speed lineup. This engine produces of 520 lbs-ft of torque. You can also go beyond this combination with the Enforcer 460 combination in either 355 or 383 cubic inch displacement. These engines use cast iron Dart cylinder heads. Peformer RPM intakes and Competition Cams valve gear.

SOURCE

American Speed Enterprises
3006 - 23rd Avenue
Moline, IL 61265

(309) 764-089

450 HP 406 thumper with Air Flow Research ported aluminum heads, 9.5:1 compression, 5.7-inch connecting rods and Competition Cams valve train.

AMERICAN SPEED ENGINE COMBINATIONS

550 HP / 502 Big Block — 4-bolt mains, cross-drilled steel crank, Chevy steel rods, forged TRW 9:1 pistons, Competition Cams hydraulic cam, Stage 1 Edelbrock ported aluminum heads, Crane roller rockers, 2.19/1.88 stainless steel valves, Edelbrock Performer RPM intake manifold, 6 quart oil pan, harmonic balancer and flexplate. Runs on pump gas with a racy idle. Likes 3.5 to 3.90 gearing and 2500 to 3000 rpm stall converters.

Brutus 555 / 434 Small Block — GM Rocket block with billet steel main caps, Stage II porter AFR aluminum cylinder heads, custom roller cam and kit, aluminum roller rockers, 4340 steel crank, 6-inch 4340 Manley steel rods, port-matched Edelbrock Victor Jr. intake, 9.5:1 compression lightweight forged pistons, 7 quart custom oil pan and steel harmonic damper. Also available with 480 to 600 horsepower depending on camshaft and application.

Brutus NOS / 406 Small Block — 4-bolt main Bow Tie block, 4340 steel crank, 6-inch 4340 steel rods, Stage II ported AFR aluminum heads, custom roller cam & kit, port-matched Edelbrock Victor Jr. intake, 9.5:1 compression lightweight forged pistons, 7 quart custom oil pan, steel harmonic damper and nitrous oxide injection system. Available up to 750 horsepower depending on camshaft and application.

Brutus 550 / 406 Small Block — 4-bolt main Bow Tie block, 4340 steel crank, 5.7-inch 4340 steel rods, ported aluminum cylinder heads, custom roller cam & kit, port-matched Edelbrock Victor Jr. intake, 9.3:1 compression lightweight forged pistons, 6 quart custom oil pan and flex plate. Available with 475 to 500 horsepower depending on the camshaft and the application.

Brutus 525 / 383 Small Block — 4-bolt main block, ported aluminum cylinder heads, port matched Edelbrock Victor Jr. intake manifold, custom roller cam and kit, roller rockers, nodular iron 383 crank, 5.7-inch rods, 9.3:1 compression light weight forged piston, 6 quart pan and a flexplate. Available from 470 to 535 horsepower with mild to rough idle depending on the camshaft.

Enforcer 500 — custom built 383 and 406 cubic inch engines with 4-bolt mains (Bow Tie block on 406), ported cast iron Dart II heads, port matched Victor Jr. intake, nodular iron crank, 5.7-inch rods, 9.3:1 compression light weight forged pistons, custom solid lifter cam and kit, roller rockers, 6 quart pan and a flexplate. Has a racy idle and likes 4.11 rear gears and 3500 rpm stall converters.

Torquer 480 EFI — 383 small block with a broad torque and power curve and racy idle. Comes with 4-bolt block, nodular crank, ported AFR aluminum heads, Crane hydraulic roller cam, roller rockers, 2.02/1.6 stainless steel valves, Melling Z/28 oil pump, 6 quart pan, Edelbrock Pro Flo EFI system, 9.5:1 compression, 5.7-inch rods and TRW forged pistons. Carries 15 inches of vacuum and likes 2500-2800 stall converters and 3.5 to 3.90 gears.

Enforcer 475 — custom built 355 cubic inch engines with 4-bolt mains, ported cast iron Dart II heads, port matched Victor Jr. intake, nodular iron crank, 5.7-inch rods, 9.0:1 compression light weight forged pistons, custom solid lifter cam and kit, Crane roller rockers, 6 quart pan and a racy idle. Likes 4.11 rear gears and 3500 rpm stall converters.

Torquer 460 & 460 EFI — 383 small block with a broad torque and power curve and racy idle. Comes with 4-bolt block, nodular crank, ported AFR aluminum heads, Competition Cams hydraulic cam, roller rockers, 2.02/1.6 stainless steel valves, Melling Z/28 oil pump, 6 quart pan, 9.5:1 compression, 5.7-inch rods and TRW forged pistons. Carries 15 inches of vacuum and likes 2500-2800 stall converters and 3.5 to 3.90 gears. Available carbureted or fuel injected.

Enforcer 460 — custom built 355 and 383 cubic inch engines with 4-bolt mains, ported cast iron Dart heads, Edelbrock Performer RPM intakes, nodular iron crank, 5.7-inch rods, 9.2:1 compression light weight forged pistons, Competition Cams cam and kit, Crane roller tipped rockers, 6 quart pan and a racy idle. These engines make 10-14 inches of vacuum and perform best with 3.90-4.11 rear gears and 3500 rpm stall converters.

Enforcer 460 — custom built 355 and 383 cubic inch engines with 4-bolt mains, ported cast iron Dart heads, Edelbrock Performer RPM intakes, nodular iron crank, 5.7-inch rods, 9.2:1 compression light weight forged pistons, Competition Cams cam and kit, Crane roller tipped rockers, 6 quart pan and a racy idle. These engines make 10-14 inches of vacuum and perform best with 3.90-4.11 rear gears and 3500 rpm stall converters.

Torquer 445, 450 and 460 — custom built 383 and 406 cubic inch engines with AFR ported aluminum cylinder heads, 2.02/1.6 stainless steel valves, Performer RPM intake manifold, new nodular iron 383 crankshaft, 5.7-inch steel rods, 9.5:1 compression ratio, lightweight forged pistons, Competition Cams camshaft and kit, This engine has a good idle and works well with all accessories, stock gearing and converters. Will have 17-18 inches of vacuum at idle and works well with 2200-2500 rpm converter and 3.4 to 3.7 gearing.

Enforcer 425 and 440 — custom built 355 and 383 cubic inch engines with cast iron ported Dart heads or ported Trick flow cylinder heads. Comes with new nodular iron crank, 9.0:1 compression forged pistons, Competition Cams cam and kit and 6 quart pan. This engine has a smooth idle and works well with all accessories, stock gearing and converters. Will have 17-18 inches of vacuum at idle and works well with 2200-2500 rpm converter and 3.7 to 4.11 gearing. These engines have a broad torque range and good mid-range power.

Torquer 400 and 425 — custom built 406 cubic inch engines with a very broad torque and power range. They feature ported Dart or Chevy heads with 2.02/1.6 valves, Edelbrock Performer RPM manifold, 5.7-inch steel rods with 9:1 forged pistons, Crane or Competition Cams cam and kit, These engines have a good (400) to crisp (425) idle and works well with all accessories. they are recommended for use with 3.08 to 3.7 gearing and 1800 to 2500 rpm converters.

Enforcer 400 TPI — custom built 355 and 383 engines with 4-bolt main blocks. They are designed for '86 and up replacement engines in TPI applications. They feature ported Dart Torquer or Trick Flow heads, Crane hydraulic cams, TRW 9:1 forged pistons, Melling Z/28 oil pump and 6 quart pan. Engines will carry 15 to 18 inches of manifold vacuum with a crisp idle and work well with all accessories. 1800 t0 2500 rpm stall converters and 3.32 to 3.90 gearing are recommended for best results. All engines come with a flexplate, or a flywheel is available at extra cost. These are all left hand dip stock engines. Left hand dip stock blocks are extra cost and injection is extra cost.

Enforcer 375 and 400 — 355, 383, 400 cubic inch engines built for maximum power and good drivability. They feature ported Dart and Trick Flow heads with 2.02/1.6 stainless valves, Performer intakes, nodular iron cranks, Crane or Cam Dynamics camshaft and kit, 9:1 forged pistons, 6 quart pan, Z/28 oil pump, and a flexplate. engines have a good to racy idle and work best with 1800-2500 rpm converters and 3.08 to 3.90 gearing.

Enforcer 350 — a 355 small block built with 9:1 forged pistons, Dart or Trick Flow heads with Stage I bowl porting, 1.94/1.5 valves, Cam Dynamics hydraulic cam and kit, Performer intake, Z/28 oil pump and 6 quart chrome oil pan. Works well with all accessories, transmissions and converters. 1800 to 2500rpm converter recommended for best performance. Likes 3.08 to 390 gearing.

Energizer 300 and 335 — Economical 9:1 355 with full blueprinting, cast iron Chevy heads, 3 angle valve job and all new parts. Engines feature Speed Pro cast pistons, Chevy steel rods, Cam Dynamics camshaft, Performer intake, Z/28 oil pump and 6 quart chrome pan. Engines have smooth idle and either 300 or 335 horsepower.

Enforcer Supercharged Engine — 355 cubic inch blueprinted and balanced engine with polished B&M 174 blower, 750 Holley carb, steel crank, Super Stock rods, Dart heads, 8:1 compression, forged pistons and dyno tested and tuned for best power. .

Enforcer Supercharged Engine — 355 cubic inch blueprinted and balanced engine with polished B&M 250 blower, Dual Holley carb, steel crank, Super Stock rods, Dart heads, 7.5:1 compression, Venolia forged blower pistons and dyno tested and tuned for best power.

Brutus Supercharged Engine — 355 cubic inch blueprinted and balanced engine with polished B&M 420 blower, Stage II ported AFR heads, 4340 steel crank, 4340 steel rods, 750 Holley carb, Dart heads, 8:1 compression forged Venolia blower pistons and dyno tested and tuned for best power.

Crate Motor Buyer's Guide

Lingenfelter Performance

Lingenfelter Performance Engineering (LPE) began as a small shop in Decatur, Indiana in the 1970's building racing engines for John Lingenfelter's C/Econo Dragster. Lingenfelter has held numerous records in various Super Stock and Competition Eliminator drag racing classes. He frequently drops out for a couple of years to pursue other racing interests, but his return is always feared by other competitors.

When GM introduced Tuned Port fuel injection in 1985, John was one of the first to see the broad potential of this new technology. Since then LPE has appplied its enormous expertise to a staggering number of EFI equipped performance cars, magazine projects and unique record holders such as the 254 mph Sledgehammer Corvette and the 298 mph Pontiac Firebird Bonneville record holder. LPE has been instrumental in developing numerous aftermarket EFI products such as the Super Ram intake marketed by Accel. These components are all put to good use on LPE customer engines.

LPE offers a complete line of small and big block Chevy crate motors to suit virtually any application including marine and towing applicataions.

All of these engine packages have been tested and thoroughly developed to ensure consistent power and quality for the crate motor consumer.

Extensive short block preparation is a hallmark of Lingenfelter Performance Engineering's engine preparation. Engines are treated to full competition style blueprinting and exacting block machining operations to ensure minimal friction losses and maximum power output.

LPE engines start with a 310 horsepower 355 small block and range all the way up to 1000 horsepower blown big blocks. In between, you can select from a full range of choices including 360 and 400 horsepower, 9:1 CR 355 small blocks designed to run on pump gas. These engines are packaged with an appropriate mix of cast and forged internals depending on the horsepower level and application.

When you step up to the 383 cubic inch small blocks you can choose either 325 horsepower with 395 lbs-ft of torque, 385 horsepower with 410 lbs-ft of torque or 475 horsepower with 448 lbs-ft of torque. All of these engines are compatible with pump gas. Two additional EFI 383 engines are also offered with 430 and 470 horsepower respectively.

LPE's Marine 502 EFI packages are available in 490 HP and 535 HP packages with an emphasis on torque. These engines use modified cast iron oval port heads, ported EFI intake, hydraulic roller cams and PF Marine exhaust manifolds.

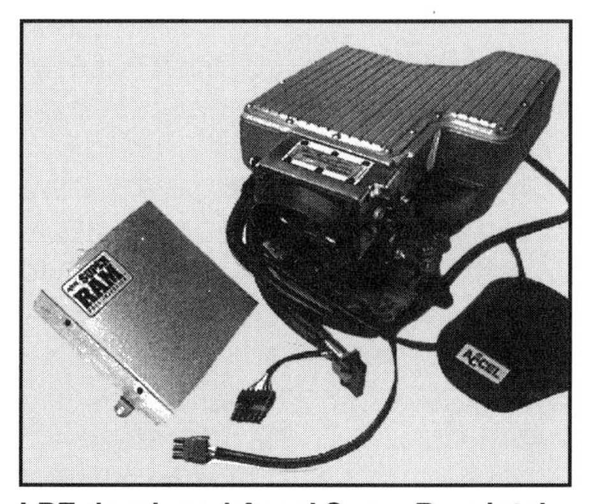

LPE developed Accel Super Ram intake is responsible for the brroad flat torque curves developed on all LPE small block EFI engines.

Here's a 420 cubic inch small block that delivers 500 HP and 550 lbs-ft of torque on 92 octane unleaded fuel. This engine makes power from 3000 to 6500 rpm. A 468 HP version is available with 512 lbs-ft of torque and a useable engine speed range of 2000 to 6000 rpm.

All cylinder blocks at Lingenfelter Performance Engineering are align honed and decked true prior to boring and honing for proper piston clearance.

Cylinder head porting is fundamental to all LPE performance engines. Depending on the application, cylinder heads are either pocket ported of fully ported to near race engine specifications.

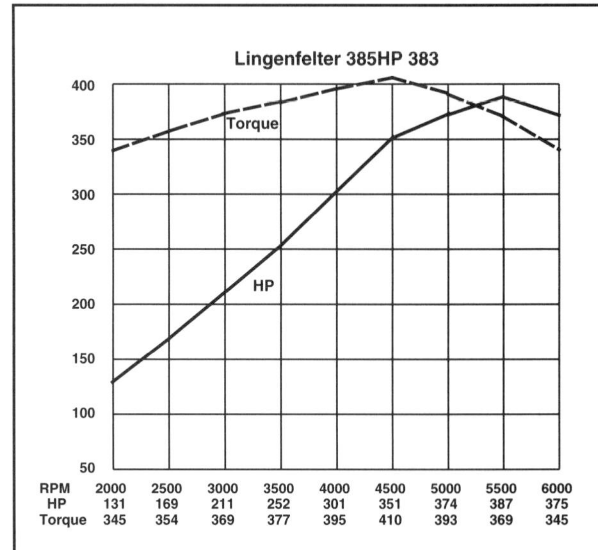

Dyno curve above is for the 385 HP LPE 383 cubic inch small block Chevy shown on the left. This engine has 9:1 compression and runs on 92 octane unleaded fuel. Note broad, flat torque curve and quick rise HP curve.

This potent 496 cubic inch big block is a popular selection because it delivers up to 612 HP and 570 lbs-ft of torque on unleaded gas. This is a hot combination for high performance Pro Street machines or Saturday night bracket racers.

The next level of performance comes with 420 cubic inch stroker small blocks. Two carbureted versions and one EFI model are available. The carbureted engines deliver either 468 horsepower and 512 lbs-ft. of torque or 500 horsepower and 550 lbs-ft. of torque. These are very hot small block street engines that are designed to remain compatible with available pump gas. The EFI version delivers 470 horsepower and 558 lbs-ft of torque and runs on pump fuel even with 11:1 CR.

Big Block Chevys

LPE big blocks are offerd in 461, 496, 502, 540 and 605 cubic inch models. Basic 461 cubic inch big blocks are available in 385, 500 and 550 horsepower versions. These are all 9:1 compression ratio engines designed to run with a single carburetor and make power in the 2000 to 6000 rpm range.

The 496 cubic inch engines deliver either 432 horsepower or 612 horsepower; one with a very smooth idle, the other with a choppy idle and enough power to tear your head off. These engines are complemented by the 502 lineup which offers 548, 664 and 720 horsepower. The 502

High end LPE performance engines typically use roller lifters, specialty aftermarket pistons and bulletproof crank and connecting rods.

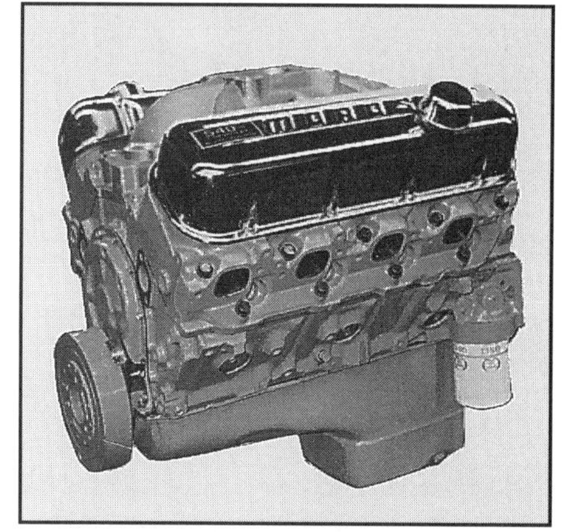
540 cubic inch carbureted big block offers 675 HP and 662 lbs-ft of torque.

selection offers up to 637 lb-ft of torque on the 720 horsepower model. All of the 502s are specified for running on unleaded pump gas.

If you step up to the 540 cubic inch line, you have 5 choices. The first is a 450 horsepower version with smooth idle and good low speed torque. Next is a 620 horsepower version with a choppy idle, but still pump gas compatible. Finally there is a 675 horsepower version with 11.5:1 compression ratio and 662 lbs-ft of torque. Two supercharged 540s are available. A twin carbureted version with 8-71 supercharger and 10 pounds of boost delivers 833 horsepower and 816 lbs-ft of torque. It runs on 104 octane unleaded fuel. A second supercharged version is also offered with electronic fuel injection and dual injectors per cylinder. This engine cranks out a staggering 1000 horsepower and 979 lbs-ft of torque.

If you like really big motors, you can also select from a pair of 605 cubic inch models; one with 590 horsepower, the other with 813 horsepower. Both engines have 11.5:1 compression ratio, The 590 horsepower version has electronic fuel injection, a smooth idle and a power range from 1600 to 5500 rpm. The 813 horsepower version has a very rough idle. It makes power from 2500 to 6500 rpm.

Many of these engines are also available in kit form for those who wish to perform their own assembly work. And if you don't see what you want, just ask. LPE can build any particular combination you want for any application.

LPE's well used engine dyno has seen literally millions of horsepower over the years. You would be surprised how many high performance aftermarket products have been developed right here in this dyno cell.

SOURCE

Lingenfelter Performance Engineering (LPE)
1557 Winchester Road
Decatur, IN 46733

(219) 724-2552
(219) 724-8761 Fax

LINGENFELTER ENGINE COMBINATIONS

355 Chevy, 310 HP, 360 lbs-ft — 9:1 compression ratio, 92 unleaded fuel, 2000-5200 rpm range, smooth idle, 2-bolt main block, cast iron blueprinted crank, Competition Cams 252 hydraulic camshaft, 5.7-inch rods, Sealed Power forged pistsons, plasma moly rings, standard volume, high pressure oil pump, Clevite 77 bearings, GM harmonic balancer, Anti pump-up lifters, true roller timing set, hardened pushrods, modified cast iron heads, 1.94/1.5 valves, Performer intake, chrome timing cover and valve covers. Requires 1-⅝-inch headers, 750 Holley, ignition & converter.

355 Chevy, 360 HP, 375 lbs-ft — 9:1 compression ratio, 92 unleaded fuel, 2000-6000 rpm range, rough idle, 2-bolt main block, cast iron blueprinted crank, Competition Cams 280 hydraulic camshaft, 5.7-inch rods, Sealed Power forged pistsons, plasma moly rings, standard volume, high pressure oil pump, Clevite 77 bearings, GM harmonic balancer, Anti pump-up lifters, true roller timing set, hardened pushrods & guide plates, modified cast iron heads, 2.02/1.6 valves, Performer intake, 1.6 ratio stamped steel rockers, chrome timing cover and valve covers. Requires 1-⅝-inch headers, 750 Holley, ignition & converter.

355 Chevy, 400 HP, 390 lbs-ft — 9:1 compression ratio, 92 unleaded fuel, 2500-6200 rpm range, choppy idle, 4-bolt main block, forged steel blueprinted crank, Competition Cams 292 hydraulic camshaft, 5.7-inch rods, Sealed Power forged pistsons, plasma moly rings, standard volume, high pressure oil pump, Clevite 77 bearings, GM harmonic balancer, Anti pump-up lifters, true roller timing set, hardened pushrods, modified D-port aluminum heads, 2.00/1.56 valves, Victor Jr. intake, chrome timing cover and valve covers. Requires 1-¾-inch headers, 750 Holley, ignition & converter.

383 Chevy, 325 HP, 395 lbs-ft — 9:1 compression ratio, 92 unleaded fuel, 2000-5200 rpm range, smooth idle, 2-bolt main block, cast iron blueprinted crank, Competition Cams 260 hydraulic camshaft, 5.565-inch rods, Sealed Power forged pistsons, plasma moly rings, standard volume, high pressure oil pump, Clevite 77 bearings, GM harmonic balancer, Anti pump-up lifters, true roller timing set, hardened pushrods, modified cast iron heads, 1.94/1.5 valves, Performer intake, chrome timing cover and valve covers. Requires 1-¾-inch headers, 750 Holley, ignition & converter.

383 Chevy, 385 HP, 410 lbs-ft — 9:1 compression ratio, 92 unleaded fuel, 2000-6000 rpm range, rough idle, 2-bolt main block, cast iron blueprinted crank, Competition Cams 280 hydraulic camshaft, 5.565-inch rods, Sealed Power forged pistsons, plasma moly rings, standard volume, high pressure oil pump, Clevite 77 bearings, GM harmonic balancer, Anti pump-up lifters, true roller timing set, hardened pushrods, modified cast iron heads, 2.02/1.6 valves, 1.6 ratio rockers, Performer intake, chrome timing cover and valve covers. Requires 1-¾-inch headers, 750 Holley, ignition & converter.

383 Chevy, 475 HP, 448 lbs-ft — 9:1 compression ratio, 92 unleaded fuel, 2500-6200 rpm range, rough idle, 4-bolt main block, cast iron blueprinted crank, Competition Cams 280 hydraulic camshaft, 5.565-inch rods, Sealed Power forged pistsons, plasma moly rings, standard volume, high pressure oil pump, Clevite 77 bearings, GM harmonic balancer, Anti pump-up lifters, true roller timing set, hardened pushrods, modified cast iron heads, 2.02/1.6 valves, 1.6 ratio rockers, Performer intake, chrome timing cover and valve covers. Requires 1-¾-inch headers, 750 Holley, ignition & converter.

420 Chevy, 468 HP, 512 lbs-ft — 11:1 compression ratio, 92 unleaded fuel, 2000-6000 rpm range, rough idle, 4-bolt main Bow Tie block, Callies 5140 steel crank, LPE 219/219 hydraulic roller camshaft, 5.850-inch Oliver billet rods, JE forged pistsons, plasma moly rings, standard volume, high pressure oil pump, Clevite 77 bearings, GM harmonic balancer, true roller timing set, hardened pushrods, modified D-port aluminum heads, 2.00/1.56 stainless steel valves, Comp Cams stainless 1.6 ratio rockers, Victor Jr. intake, chrome timing cover and valve covers. Requires 1-¾-inch headers, 750 Holley, ignition & converter.

420 Chevy, 500 HP, 550 lbs-ft — 11:1 compression ratio, 92 unleaded fuel, 2000-6000 rpm range, rough idle, 4-bolt main Bow Tie block, Callies 5140 steel crank, LPE 219/219 hydraulic roller camshaft, 5.850-inch Oliver billet rods, JE forged pistsons, plasma moly rings, standard volume, high pressure oil pump, Clevite 77 bearings, GM harmonic balancer, true roller timing set, hardened pushrods, modified D-port aluminum heads, 2.00/1.56 stainless steel valves, Comp Cams stainless 1.6 ratio rockers, Victor Jr. intake, chrome timing cover and valve covers. Requires 1-¾-inch headers, 750 Holley, ignition & converter.

461 Chevy, 385 HP, 485 lbs-ft — 9:1 compression ratio, 92 unleaded fuel, 2000-5200 rpm range, smooth idle, 2-bolt main block, cast iron 454 crank, Comp Cams 260 hydraulic camshaft, ⅜-inch bolt rods, Speed Pro forged pistsons, plasma moly rings, standard volume, high pressure oil pump, Clevite 77 bearings, GM harmonic balancer, Anti pump-up lifters, true roller timing set, hardened pushrods, modified cast iron heads, 2.065/1.720 stainless steel valves, 1.7 ratio stamped steel rockers, Performer intake, timing cover and valve covers. Requires 1-⅞-inch headers, 850 Holley, ignition & converter.

461 Chevy, 500 HP, 500 lbs-ft — 9:1 compression ratio, 92 unleaded fuel, 2000-6000 rpm range, choppy idle, 2-bolt main block, cast iron 454 crank, Comp Cams 280 hydraulic camshaft, ⅜-inch bolt rods, Speed Pro forged pistsons, plasma moly rings, standard volume, high pressure oil pump, Clevite 77 bearings, GM harmonic balancer, Anti pump-up lifters, true roller timing set, hardened pushrods, modified cast iron heads, 2.25/1.88 stainless steel valves, 1.7 ratio Comp Cams stainless steel rockers, Dart aluminum intake, timing cover and valve covers. Requires 1-⅞-inch headers, 850 Holley, ignition & converter.

461 Chevy, 550 HP, 500 lbs-ft — 9:1 compression ratio, 92 unleaded fuel, 2000-6000 rpm range, choppy idle, 2-bolt main block, cast iron 454 crank, Comp Cams 260 solid lifter camshaft, ⅜-inch bolt rods, Speed Pro forged pistsons, plasma moly rings, standard volume, high pressure oil pump, Clevite 77 bearings, GM harmonic balancer, Anti pump-up lifters, true roller timing set, hardened pushrods, modified cast iron heads, 2.065/1.720 stainless steel valves, 1.7 ratio stamped steel rockers, Dart aluminum intake, timing cover and valve covers. Requires 1-⅞-inch headers, 850 Holley, ignition & converter.

496 Chevy, 432 HP, 561 lbs-ft — 9.2:1 compression ratio, 92 unleaded fuel, 1600-5000 rpm range, smooth idle, 2-bolt main block, Callies 4340 steel crank, Comp Cams 252 hydraulic camshaft, ⅜-inch bolt rods, Speed Pro forged pistsons, plasma moly rings, standard volume, high pressure oil pump, Clevite 77 bearings, GM harmonic balancer, Anti pump-up lifters, true roller timing set, hardened pushrods, modified cast iron heads, 2.065/1.720 stainless steel valves, 1.7 ratio stamped steel rockers, Performer intake, timing cover and valve covers. Requires 1-⅞-inch headers, 850 Holley, ignition & converter.

496 Chevy, 612 HP, 561 lbs-ft — 9.2:1 compression ratio, 92 unleaded fuel, 2000-6000 rpm range, choppy idle, 2-bolt main block, Callies 4340 steel crank, Comp Cams CB288AR10 roller camshaft, ⅜-inch bolt rods, Speed Pro forged pistsons, plasma moly rings, standard volume, high pressure oil pump, Clevite 77 bearings, GM harmonic balancer, Anti pump-up lifters, true roller timing set, hardened pushrods, modified cast iron heads, 2.25/1.88 stainless steel valves, 1.7 ratio Comp Cams stainless steel rockers, modified Dart aluminum intake, timing cover and valve covers. Requires 1-⅞-inch headers, 850 Holley, ignition & converter.

502 Chevy, 548 HP, 555 lbs-ft — 9:1 compression ratio, 92 unleaded fuel, 2000-5200 rpm range, smooth idle, 4-bolt main Gen V block, 1053 GEN V steel crank, Comp Cams 286 hydraulic camshaft, ⅜-inch bolt rods, forged JE pistsons, plasma moly rings, standard volume, high pressure oil pump, Clevite 77 bearings, GM harmonic balancer, Anti pump-up lifters, true roller timing set, hardened pushrods, modified cast iron heads, 2.25/1.88 stainless steel valves, 1.7 ratio stamped steel rockers, Performer intake, timing cover and valve covers. Requires 1-⅞-inch headers, 850 Holley, ignition & converter.

502 Chevy, 664 HP, 605 lbs-ft — 9:1 compression ratio, 92 unleaded fuel, 2000-6500 rpm range, rough idle, 4-bolt main Gen V block, 1053 GEN V steel crank, Comp Cams 296 mechanical roller camshaft, 7/16-inch bolt rods, forged JE pistsons, plasma moly rings, standard volume, high pressure oil pump, Clevite 77 bearings, GM harmonic balancer, Anti pump-up lifters, true roller timing set, hardened pushrods, modified cast iron heads, 2.25/1.88 stainless steel valves, 1.7 ratio Comp Cams stainless steel rockers, Dart aluminum intake, timing cover and valve covers. Requires 2-⅛-inch headers, 850 Holley, ignition & converter.

502 Chevy, 720 HP, 637 lbs-ft — 11:1 compression ratio, 92 unleaded fuel, 2000-5500 rpm range, rough idle, 4-bolt main Gen V block, 1053 GEN V steel crank, Comp Cams 296 mechanical roller camshaft, 7/16-inch bolt rods, forged JE pistsons, plasma moly rings, standard volume, high pressure oil pump, Clevite 77 bearings, GM harmonic balancer, Anti pump-up lifters, true roller timing set, hardened pushrods, Brodix aluminum heads, 2.25/1.88 stainless steel valves, 1.7 ratio Comp Cams stainless steel rockers, Dart aluminum intake, timing cover and valve covers. Requires 2-¼-inch headers, 1050 Holley, ignition & converter.

540 Chevy, 450 HP, 590 lbs-ft — 9.2:1 compression ratio, 92 unleaded fuel, 1600-5500 rpm range, smooth idle, 4-bolt main Bow Tie block, 4340 Callies steel crank, Comp Cams 252 hydraulic camshaft, ⅜-inch bolt rods, forged JE pistsons, plasma moly rings, standard volume, high pressure oil pump, Clevite 77 bearings, GM harmonic balancer, Anti pump-up lifters, true roller timing set, hardened pushrods, modified cast iron heads, 2.065/1.720, stainless steel valves, 1.7 ratio stamped steel rockers, Performer intake, timing cover and valve covers. Requires 1-⅞-inch headers, 850 Holley, ignition & converter.

540 Chevy, 620 HP, 630 lbs-ft — 9.2:1 compression ratio, 92 unleaded fuel, 2500-6500 rpm range, choppy idle, 4-bolt main Bow Tie block, 4340 Callies steel crank, Comp Cams 288AR10 roller camshaft, ⅜-inch bolt rods, forged JE pistsons, plasma moly rings, standard volume, high pressure oil pump, Clevite 77 bearings, GM harmonic balancer, true roller timing set, hardened pushrods, modified cast iron heads, 2.25/1.88, stainless steel valves, 1.7 ratio stainless steel rockers, modified Dart intake, timing cover and valve covers. Requires 2-⅛-inch headers, 850 Holley, ignition & converter.

540 Chevy, 675 HP, 662 lbs-ft — 11.5:1 compression ratio, 92 unleaded fuel, 2500-6500 rpm range, choppy idle, 4-bolt main Bow Tie block, 4340 Callies steel crank, Comp Cams 288AR10 roller camshaft, 7/16-inch bolt rods, forged JE pistsons, plasma moly rings, standard volume, high pressure oil pump, Clevite 77 bearings, GM harmonic balancer, true roller timing set, hardened pushrods, modified cast iron heads, 2.25/1.88, stainless steel valves, 1.7 ratio stainless steel rockers, modified Dart intake, timing cover and valve covers. Requires 2-⅛-inch headers, 850 Holley, ignition & converter.

540 Chevy, 833 or 1000 HP, 816 or 979 lbs-ft, Supercharged — 7.5:1 compression ratio, 104 unleaded fuel, 2000-6500 rpm range, choppy idle, 4-bolt Bow Tie block, 4340 Callies steel crank, roller camshaft, Oliver billet rods, forged JE pistsons, plasma moly rings, modified Brodix aluminum heads, 2.30/1.88, stainless steel valves, 1.7 ratio stamped steel rockers, modified Dart intake, 8-71 Supercahrger, water/air intercooler, pulley and drive kit, timing cover and valve covers. Requires 2-¼-inch headers, 850 Holley, ignition & converter.

Callaway Engines

The Callaway SuperNatural® 383 Chevy small block V8 is based on the Chevy LT1 V8 and can be configured to deliver up to 500 horsepower if desired. These engines are based on the same combination used in the Callaway SuperNatural Corvette Le Mans and the new Callaway C& Sports GT. Callaway doesn't offer a broad range of crate motors, but rather a focused number of 383 engines designed for use in Corvettes, Camaros, Impala SS and street rods. these 383 cubic inch engines offer great performance, the drivability of a stock LT1, reliability and simple maintenance.

The engines begin with clean new cylinder blocks that are fitted with a 3.75-inch stroke forged 5140 steel crankshaft. High silicon forged aluminum pistons and low friction, file-fit rings are teamed with 4340 forged steel connecting rods to form a high performance foundation. These efforts are directed at making a bulletproof lower end to handle the power levels generated by the induction components. To make the power, Callaway engine builders start with a new set of LT1 aluminum cylinder heads. Revised valve stem angles and reshaped combustion chambers unshroud the periphery of the valve heads and distribute the intake charge more uniformly for efficient combustion and detonation resistance. The intake and exhaust ports have proprietary shaped developed during testing

High silicon forged aluminum pistons are installed with file-fit low friction piston rings. Cylinder bores are carefully finished to ensure the best possible ring seal.

for the Callaway GT2 international racing engines. The heads are optimized with 1.6:1 ration roller tipped rocker arms; reducing the need for bigger cam profiles which may be less driveable. The valve springs are matched to the valve train loads and secured with lightweight retainers and keepers.

Depending on the application, some versions receive a large-volume modified plenum to improve low and mid-range performance. Whether stock or modified, the plenum is carefully ported and matched to the new cylinder heads and the throttle body size is increased to either 52mm or 58mm.

High quality internals include these forge pistons and extra heavy duty connecting rods to ensure total reliability.

The exhaust side incorporates stainless steel Tri-Y headers with 1.75-inch primaries and heavy duty flanges. The collectors are sized at 2.5-inches.

Delco Electronics powertrain managements systems as installed on the 92-93 Corvette are used to deliver great power, responsiveness and the kind of smooth performance that many enthusiasts are looking for in a street engine. Callaway will program the entire powertrain including the new electronically shifted 4L60E and 4L80E transmissions to get the most from each engine based on your application, drivetrain and intended usage.

Callaway stresses that these engines are built for longevity as well as performance. Electronic engine management and electronic ignition reduces tuning needs and increases spark plug life. Callway's engines are a little different from other crate motor engines because they stress the high end of the spectrum. they aren't cheap, but you get what you pay for.

They also believe that looks are important. This includes the quality, fit and finish of all external components. Callaway will custom paint, plate or anodize engine components to suit your tastes, Custom machined brackets, dress covers, dry sump systems and the latest carbon fiber accessories are all available from Callaway.

Polished aluminum valve covers and unique Callaway logo carbon fiber injector covers provide a unique look for each Callaway engine. They recommend the use of these engines for high end street rods and special application Corvettes, Camaros, Firebirds, Impalas and other GM performance cars. Moreover, the Callaway 383 V8 engines are fully emissions compliant for most applications.

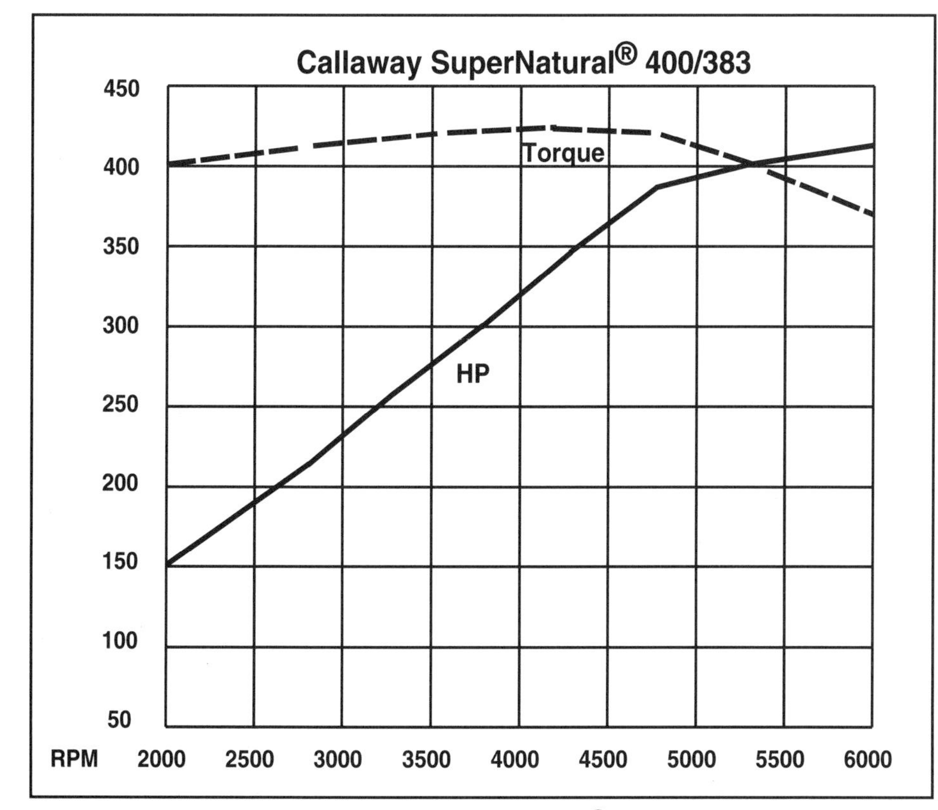

Typical power curve for a Callaway SuperNatural® 400/383 shows the very flat torque curve and quick rise horsepower curve available from these engines.

Jasper Engines

Jasper Performance Products is a major engine and transmission supplier for industrial equipment and automotive engines. Located in Jasper, Indiana, Jasper Engines offer a variety of high performance Chevy and Ford V8s for use in high performance street rods and street machines, towing packages and other applications.

The first engine we'll discuss is their Class II version Chevy 350 rated at 328 horsepower and 383 lbs-ft. of torque. This engine is offered for the serious performance minded enthusiast who needs power for a specific application. This engine is built with heavy duty performance in mind, hence special components and procedures are selected and matched to obtain maximum performance. Jasper specifies this engine for performance cars and trucks, street rods and restored muscle cars using small to medium sized engines where good low speed torque is important. This engine is also specified for use as an OEM replacement engine in late model vehicles.

The engine is built with a 8.75:1 compression ratio. It includes high performance rod bolts and nuts, high rev lifters and performance valve springs with retainers, high volume oil pump, double roller timing chain, flat top hypereutectic pistons with moly rings, harmonic balancer, screw-in rocker arm studs, high performance stainless steel 1.94-inch intake valves and 1.5-inch exhaust valves, Class II camshaft, high performance bearings, head gaskets, 4-bolt main block and new Dart Street Replacement cylinder heads.

Engine options available with these engines include magnum roller tipped rocker arms, aluminum roller rocker arms, Edelbrock Performer RPM intake installed, Performer intake installed and a steel crankshaft.

Jasper engines are carefully pre-

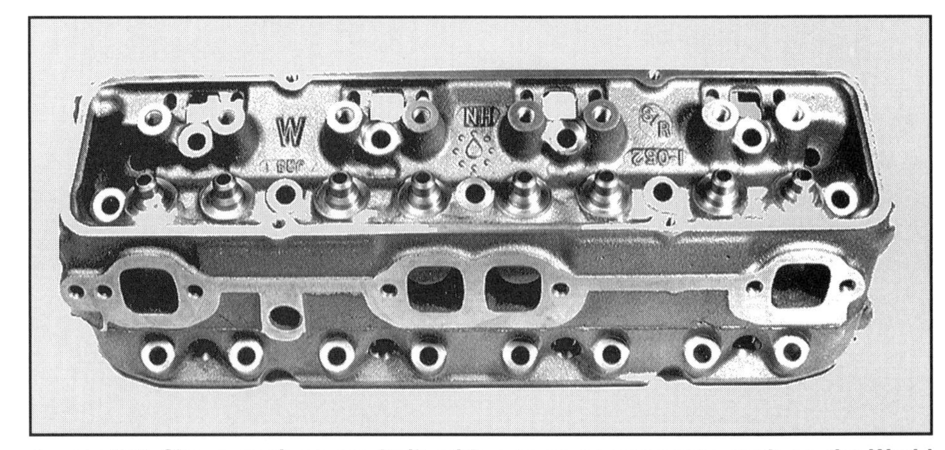

Jasper 350 Chevy engines are built with proven components such as the World Products S/R Street Replacement cylinder head out fitted with premium stainless steel valves and high performance valve gear.

Jasper Class II 350 V8 delivers 350 horsepower at 5500 rpm and 389 lbs-ft. of torque at 4200 rpm. It is recommended for street rods and street machines or applications requiring strong mid-range torque and power.

pared to the same exacting standards as their industrial engines. The blocks are pressure cleaned in a Kolene salt bath, decks are squared with precise surface finish specs, main saddles are align honed, bores are computer bored and torque plate honed, the crankshaft is precision machined with oil hole chamfering and polishing and each engine is live tested.

The Class II hydraulic camshaft used in these engines features 214° intake and 224° exhaust duration at .050-inch. Valve lift is .443-inch on the intakes and .465 on the exhaust side. The lobe centers are 107° intake and 117° exhaust.

These specs provide a fair idle with mild lope. The operating range is from 2000 rpm to 4800 rpm with good mid-range torque in the 2400 to 3200 rpm range. Jasper claims good fuel economy and good towing capacity with the correct rear axle ratio. They also suggest that this engine performs very well for mild bracket racing applications. It is compatible with both automatic and manual transmissions and the recommended axle ratio is 3.70:1. A stock or aftermarket 2 plane intake manifold is recommended with either a two barrel carburetor or a small four barrel carburetor. Headers, manifold and carburetor are not included with the engine, but are available for additional cost.

350 HP 350

The 350 horsepower 350 cubic inch Chevy small block is built with all the same specs and procedures as the 328 horsepower 350 except that it has 9:1 compression ratio

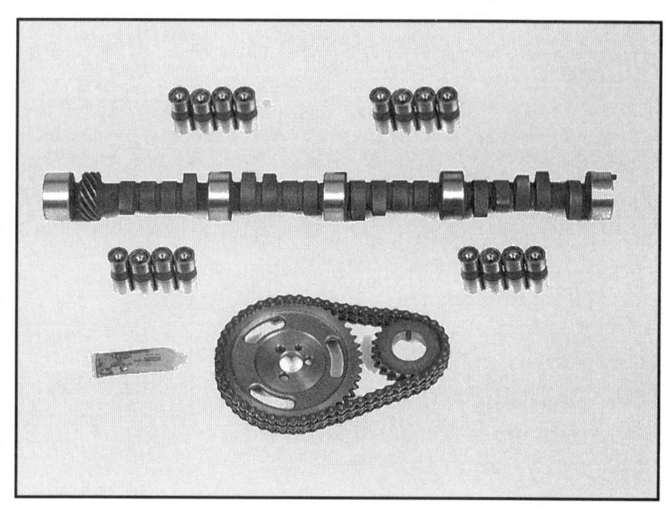

The 350 horsepower version of the Class II Chevy V8 gets its manners from a high performance hydraulic camshaft that offers 230° duration at .050-inch valve lift, .480-inch valve lift and 107° intake lobe centerline and 111° degree exhaust centerline.

Jasper small block Chevys are available with a full list of optional equipment for the induction system, ignition system, engine dress and accessory packages. They will also dyno test your engine and provide a printout for added cost.

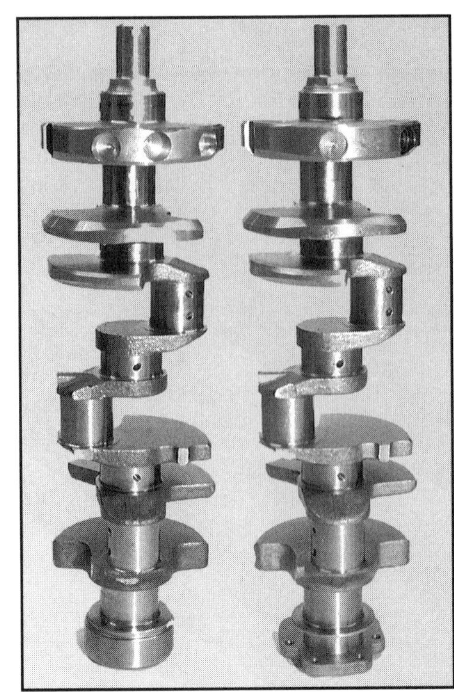

Chevy engines have the option of either a forged or cast crankshaft. Late model engines require the use of a one-piece rear seal crankshaft (left).

and a hotter camshaft. This engine also uses brand new Dart Street Replacement heads with 2.02/1.6-inch stainless steel valves.

The camshaft in this engine features 230 degrees duration at .050-inch valve lift with an advertised duration of 292 degrees. Valve lift is .480-inch on both the intake and the exhaust valve. This cam has a pretty good lope and reduced idle vacuum, but it makes excellent power.

FORD 302 & 351W

Jasper's small block Ford engines are offered in a variety of horsepower levels. Both the 302 and the 351W have three different version plus the optional versions for use in Cobra kit cars.

All engines are built with computerized boring, torque plate honing, squared decks, align honing and option choices for Dart Windsor or Trick Flow Specialty cylinder heads. Headers, manifold and carburetor are not included with these engines.

The 280 horsepower 302 uses a flat tappet camshaft with 214/224 degrees duration and .472/.492-inch valve lift. This engine makes power to 5100 rpm.

The next level for the 302 is 302 horsepower with a hydraulic roller camshaft measuring 222/232 degrees at .050-inch lift and .510/.534-inch lift on the intake and exhaust valves respectively. The 302 horsepower engine revs to 5500 rpm.

The top of the line 302 is a 350 horsepower version that runs to 6500 rpm with a flat tappet camshaft measuring 248/248 degrees at .050-inch valve lift and .560/.560-inch lift.

In the 351W line Jasper starts with a flat tappet cam version that makes 305 horsepower at 4900 rpm. The cam features 214/224

Small block Ford engines are built with precision machining and high quality parts. A long list of options and extras accompanies each engine. Kit Car enthusiasts can also choose dress-up items, accessories and even transmissions to go with their engine purchase.

degrees duration at .050-inch lift and .472/.496-inch valve lift.

The next level is a flat tappet cam 351W that delivers a strong 365 horsepower at 5800 rpm. This engine has a moderately big cam and a lopy idle. The cam measures 224/234 at .050-inch lift and .496/.520 lift.

The top of the line stoutest version of the 351W makes 405 horsepower at 5700 rpm using a big flat tappet camshaft with 234/244 degree duration at .050 and .520/.544 valve lift.

All of these engines can be ordered with Dart Windsor or Trick Flow Specialties cylinder heads that will increase power output. Induction options include a 650 CFM or 750 CFM Holley double pumper four barrel, chrome fuel line with pressure gauge, Holley high volume fuel pump, ARP stainless steel intake bolts and either a Performer or Performer RPM intake manifold.

The Cobra Kit Cars versions also offer optional Mallory Unilite distributors, Taylor or Moroso ignition wires, Chrome engine accessory and dress up package, polished aluminum water pump and alternator drive pulleys and dyno testing if desired. Like most major crate motor suppliers, Jasper takes great pains to ensure that their crate motors are built with the highest quality and thoroughly tested to verify power levels.

One of Jasper's more interesting engine series is the 302 & 351W Cobra Kit Car Package. These are 9:1 engines with either forged or hypereutectic pistons, a Wolverine Blue Racer camshaft. Induction system choices are optional.

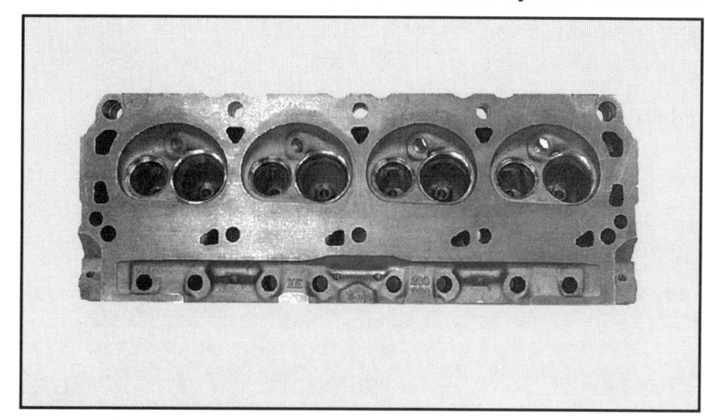

Ford 302 & 351W Cobra Kit Car engines use stock Ford cylinder heads with 64 cc combustion chambers, new valves, performance valve springs, chrome moly retainers and hardened keepers.

Ford short blocks are built with either forged or hypereutectic pistons and a compression ratio of 9:1. Rods are fitted with high performance ARP rod bolts Six different power levels are available based on camshaft selection. Engines start at 280 HP and run up to 405 HP.

SOURCE

Jasper Engines
815 Wernsing Rd.
P. O. Box 650
Jasper, IN 47547-0650

(800) 827-7455
(812) 634-1820 Fax

JASPER ENGINE COMBINATIONS

Chevy Small Blocks

350 Chevy, 328 HP, 383 lbs-ft — 8.75:1 compression ratio, high performance rod bolts and nuts, high rev lifters and performance valve springs with retainers, high volume oil pump, double roller timing chain, flat top hypereutectic pistons, moly rings, harmonic balancer, screw-in rocker arm studs, high performance stainless steel 1.94/1.5 valves, high performance bearings, high performance head gaskets, 4-bolt main block, Dart S/R cylinder heads, Class II camshaft with 214/224 duration at .050, .443/.465 valve lift, optional roller tipped rocker arms, aluminum roller rockers, Performer intake or Performer RPM intake installed, steel crankshaft, recommended 3.70 gearing, compatible with stock type auto transmission converters.

350 Chevy, 350 HP, 389 lbs-ft — 9:1 compression ratio, high performance rod bolts and nuts, high rev lifters and performance valve springs with retainers, high volume oil pump, double roller timing chain, flat top hypereutectic pistons, moly rings, harmonic balancer, screw-in rocker arm studs, high performance stainless steel 2.02/1.6 valves, high performance bearings, high performance head gaskets, 4-bolt main block, Dart S/R cylinder heads, Class II camshaft with 230/230 duration at .050, .480/.480 valve lift, optional roller tipped rocker arms, aluminum roller rockers, Performer intake or Performer RPM intake installed, steel crankshaft, recommended 3.70 gearing, compatible with stock type auto transmission converters. 750 Holley carburetor recommended.

Induction Options — Holley 750 CFM or 650 CFM four barrel, double pumper with electric choke, polished aluminum, mechanical secondaries, chrome fuel inlet line with pressure gauge, steel braided fuel line, Holley high volume fuel pump, ARP stainless steel intake bolts, Edelbrock Victor Jr. or Performer intake installed.

Ignition Options — Accel electronic ignition, vacuum advance distributor, Accel super duty coil, coil bracket, ballast resistor, Taylor or Moroso spark plug wires and spark plugs.

MSD Ignition System — MSD HEI kit, Taylor ignition wires, 90 degree, spark plugs, price with labor to install.

Miscellaneous Accessories — chrome thermostat outlet and gasket, oil dipstick and tube, chrome breather with PCV valve, chrome breather, miscellaneous bolts, gaskets and brackets.

Front of Engine Dress — New water pump, polished aluminum single groove pulleys for water pump, damper and alternator, multi-groove pulleys available at extra cost, alternator bracket, chrome 60 amp alternator and belt.

Dyno Testing — complete test with printout sheet, extra cost

Ford Small Blocks

302 Ford, 280 HP — 8.75:1 compression ratio, high performance rod bolts and nuts, high rev lifters and performance valve springs with retainers, high volume oil pump, double roller timing chain, flat top hypereutectic pistons, moly rings, harmonic balancer, screw-in rocker arm studs, high performance stainless steel valves, high performance bearings, high performance head gaskets, 2-bolt main block, stock heads, Class II flat tappet camshaft with 214/224 degrees at .050, .475/.496 valve lift, optional roller tipped rocker arms, aluminum roller rockers, Performer RPM intake installed, steel crankshaft, recommended 3.70 gearing, compatible with stock type auto transmission converters.

302 Ford, 302 HP — 8.75:1 compression ratio, high performance rod bolts and nuts, high rev lifters and performance valve springs with retainers, high volume oil pump, double roller timing chain, flat top hypereutectic pistons, moly rings, harmonic balancer, screw-in rocker arm studs, high performance stainless steel valves, high performance bearings, high performance head gaskets, 2-bolt main block, stock heads, Class II hydraulic roller camshaft with 222/232 degrees duration, .510/.534 valve lift, optional roller tipped rocker arms, aluminum roller rockers, Performer RPM intake installed, steel crankshaft, recommended 3.70 gearing, compatible with stock type auto transmission converters.

302 Ford, 350 HP — 9:1 compression ratio, high performance rod bolts and nuts, high rev lifters and performance valve springs with retainers, high volume oil pump, double roller timing chain, flat top hypereutectic pistons, moly rings, harmonic balancer, screw-in rocker arm studs, high performance stainless steel valves, high performance bearings, high performance head gaskets, 2-bolt main block, stock heads, Class II flat tappet camshaft with 248/248 degrees duration, .560/.560 valve lift, optional roller tipped rocker arms, aluminum roller rockers, Performer RPM intake installed, steel crankshaft, recommended 3.70 gearing, compatible with stock type auto transmission converters.

351W Ford, 305 HP — 8.75:1 compression ratio, high performance rod bolts and nuts, high rev lifters and performance valve springs with retainers, high volume oil pump, double roller timing chain, flat top hypereutectic pistons, moly rings, harmonic balancer, screw-in rocker arm studs, high performance stainless steel valves, high performance bearings, high performance head gaskets, 2-bolt main block, stock heads, Class II flat tappet camshaft with 214/224 degrees duration, .472/.496 valve lift, optional roller tipped rocker arms, aluminum roller rockers, Performer RPM intake installed, steel crankshaft, recommended 3.70 gearing, compatible with stock type auto transmission converters.

351W Ford, 365 HP — 9:1 compression ratio, high performance rod bolts and nuts, high rev lifters and performance valve springs with retainers, high volume oil pump, double roller timing chain, flat top hypereutectic pistons, moly rings, harmonic balancer, screw-in rocker arm studs, high performance stainless steel valves, high performance bearings, high performance head gaskets, 2-bolt main block, stock heads, Class II flat tappet camshaft with 224/234 degrees duration, .492/.520 valve lift, optional roller tipped rocker arms, aluminum roller rockers, Performer RPM intake installed, steel crankshaft, recommended 3.70 gearing, compatible with stock type auto transmission converters.

351W Ford, 405 HP — 9:1 compression ratio, high performance rod bolts and nuts, high rev lifters and performance valve springs with retainers, high volume oil pump, double roller timing chain, flat top hypereutectic pistons, moly rings, harmonic balancer, screw-in rocker arm studs, high performance stainless steel valves, high performance bearings, high performance head gaskets, 2-bolt main block, stock heads, Class II flat tappet camshaft with 234/244 degrees duration, .520/.544 valve lift, optional roller tipped rocker arms, aluminum roller rockers, Performer RPM intake installed, steel crankshaft, recommended 3.70 gearing, compatible with stock type auto transmission converters.

Cobra Kit Car Package Options

Ignition System — Mallory Unilite electronic ignition, vacuum advance distributor, Mallory chrome super duty coil, Coil brackets, ballast resistor, Taylor or Moroso spark plug wires and spark plugs.

Cobra Dress-up — "Cobra POwered By Ford" valve covers, Cobra oval air cleaner, Cobra T-shaped aluminum oil pan, chrome thermostat housing and gasket, oil dipstick and tube, breather with PCV valve, chrome breather, bolts gaskets and brackets.

Accessories — Water pump with polished double groove pulleys for water pump, damper and alternator, 60 amp alternator, bracket and belt.

Cobra Transmissions & Accessories — Lakewood or McLeod bellhousing, Hays 30 lb. flywheel, Hays or Zoom street performance clutch assembly, hydraulic clutch, mechanical clutch, Cable clutch, street 5-speed, Tremac 5-speed, T-10 4 speed, Borg-Warner T-5, Toploader, C4 automatic or C6 automatic.

Scoggin-Dickey Engines

Scoggin-Dickey has been one of the nation's major mail-order engine suppliers for several decades. It is a full service parts center featuring GM engines and a complete line of mail-order aftermarket high performance parts. They specialize in Goodwrench engines, long block and short block assemblies,Chevy HO replacement engines and a number of variations that are built in-house.

The standard 350 special is specified for replacement use in any 1969-85 Chevy. It delivers 250 horsepower and comes with a 4-bolt main cylinder block and a 36 month/50,000 mile warranty. These are factory assembled long block engines including valve covers and oil pan. You will have to add an intake manifold, fuel system, ignition system, water pump, balancer and flywheel.

Also included in their lineup of factory engines are 305 TBI replacements for 1987 to 1991 Tuned Port Injection cars and 350 TBI engines for the same years. Scoggin-Dickey can also supply Goodwrench replacement engines for virtually any GM application including trucks and diesels.

Other crate motor applications

Basic replacement crate engine 350 delivers 250 horsepower with 4-bolt main block and a full 36 month, 50,000 mile warranty.

295 horsepower and 330 horsepower 350 V8s are offered with aluminum cylinder heads and strong, but streetable camshafts to provide solid high performance engine swap material.

one -stop shopping parts center. You wouldn't want to put a lot of nitrous oxide or supercharger boost through these engines, but they are the perfect choice for warmed over street machines, street rods, pickups, RV's, low riders, etc.

If you happen to favor a particularly unique combination of intake manifold, carburetor, ignition, exhaust, or even camshaft, you can easily reconfigure either of these engines to suit you own personal tastes. Check with your Scoggin-Dickey salesman if you are going to change the cam because it may void the warranty. Beyond that you can use these engines pretty hard and expect to receive strong, reliable performance with decent fuel economy.

include 295 horsepower and 330 horsepower Scoggin-Dickey performance engines with aluminum cylinder heads. These engines are fitted with special camshafts and aluminum cylinder heads but they do not include intake, distributor, water pump, flexplate or balancer.

The 295 horsepower version comes with a hydraulic camshaft featuring 194°/202° duration at .050-inch lift and .390/.410-inch valve lift. The recommended combination for finishing off this engine includes an Edelbrock Performer intake manifold, Edelbrock Performer carburetor, GM HEI distributor, Hedman headers, MSD plug wires, Edelbrock water pump and March pulleys; all of which are available from Scoggin-Dickey.

The 330 horsepower version has a hotter camshaft with 214°/224° duration at .050-inch lift and .442/.465-inch valve lift. The same aftermarket parts mentioned above are also recommended for this engine.

Both of these engines are based on high quality, factory assembled short blocks. Either engine could be hopped up even more with a more aggressive mix of high performance parts, all of which are available from Scoggin-Dickey's

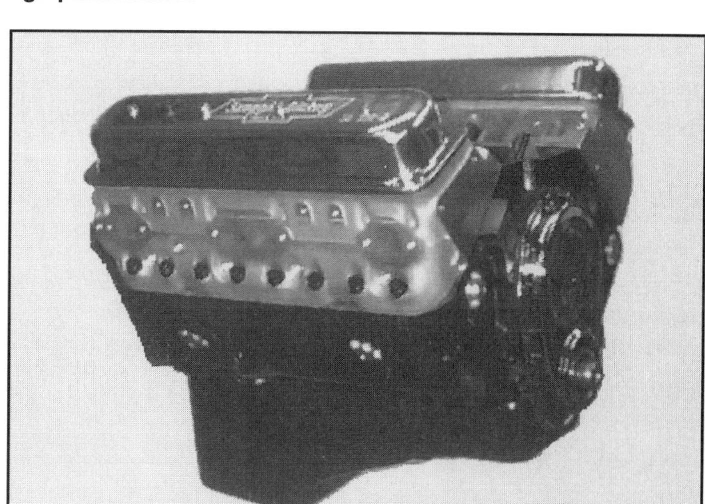

An aluminum head Scoggin-Dickey 350 Chevy small block dressed out with custom Scoggin-Dickey polished valve covers.

Scoggin-Dickey 350 Chevy aluminum head small block equipped with electronic fuel injection and Edelbrock custom polished valve covers.

Scoggin-Dickey 345 horsepower Chevy 350 HO crate motor is also a popular choice for high performance engine swaps, engine replacements and other special applications where strong reliable power is required

300 horsepower 350 SP engine is a 9:1 compression Chevy small block that will accept a variety of aftermarket speed equipment. This engine comes equipped with chrome valve covers and timing cover, but does not have an intake manifold, balancer, ignition, carburetor or exhaust manifolds.

If you call the Scoggin-Dickey Performance Hotline at 1-800-456-0211, you can purchase a copy of the Goodwrench engine catalog which details all the new and remanufactured GM engines available. Scoggin-Dickey sells all of these engines and this catalog will help you make the right choice.

Big Block Chevys

Big block Chevys are among the most popular combinations sought by high performance enthusiasts. Scoggin-Dickey offers a partial 502 assembly under PN 10185059. It features a marine standard deck height iron block with 4-bolt mains, a 4.00-inch stroke forged steel crank and forged pistons. the compression ratio is 8.75:1 with 118cc combustion chambers.The crank is 4340 forged steel and the rod bolts are heavy duty 7/16-inch units. The engine includes block, crankshaft, rods, pistons, flexplate, damper, oil pump, pan, and front cover. Cylinder heads and camshaft are not included. This is a Mark V version of the 502, but GEN VI versions will be available in mid-96.

The complete HO 502 engine assembly PN 10185085 is a high torque engine offering exceptional performance value in an over-the counter-package. This engine has been rated up to 440 horsepower at 5250 rpm and 515 lbs-ft of torque at 3500 rpm.

The short block is the same as the standard short block assembly, but it is fitted with high performance rectangular cylinder heads, 2.19/1.88 valves had heavy duty valve gear. The camshaft is a hydraulic unit with .500-inch valve lift and 220° duration at .050-inch lift. The standard 502 uses non-adjustable valve train, while the new GEN VI brings back the adjustability of the old Mark IV design. Scoggin-Dickey sells replacement pieces to convert the Mark IV valvetrain over for use on later model GEN II and GEN VI performance engines.

Long block 502 assembly provides one of the most powerful and reliable crate engines available. Scoggin-Dickey has the Mark V versions available and will have the GEN VI in mid-96.

Partial or short block 502 engine assemblies are a good way to go for those who may wish to run their own cylinder heads. The 502 short blocks comes alive quite well when fitted with World Products Merlin oval port cylinder heads.

SOURCE

SCOGGIN-DICKEY
5901 Spur 327
Lubbock, TX 79424

Order Line (800) 456-0211
Tech Line (806) 798-4108

Electromotive, Inc.

Electromotive is a fast growing company specializing in high performance engine management systems with integrated electronic fuel injection and high energy distributorless ignition systems. Their systems include PC based engine monitoring and control which accommodates both speed density and mass air flow engine management.

The unique abilities of Electromotive's highly tunable system prompted the company to begin offering crate engines equipped with its high performance TEC II Total Engine Control EFI and direct ignition systems. These packages offer fully programmable spark curves and sequential fuel injection in a totally integrated crate motor package.

Electromotive's crate motor program extends to 350 cubic inch small block Chevys, 454 and 502 big block Chevys and 5.0L small block Fords. The same engine management componentry used to control these engines can also be used to manage any other size GM or Ford engine and Mopars if desired. The management systems are also sold separately for application to your unique engine application—a 383 or 406 small block Chevy for example.

The Tuned Port Injection kit used on small block Chevy crate motors incorporates TEC-II engine management with direct ignition and closed loop fuel control, intake manifold with air door, plenum and runners, wiring harness, fuel injec-

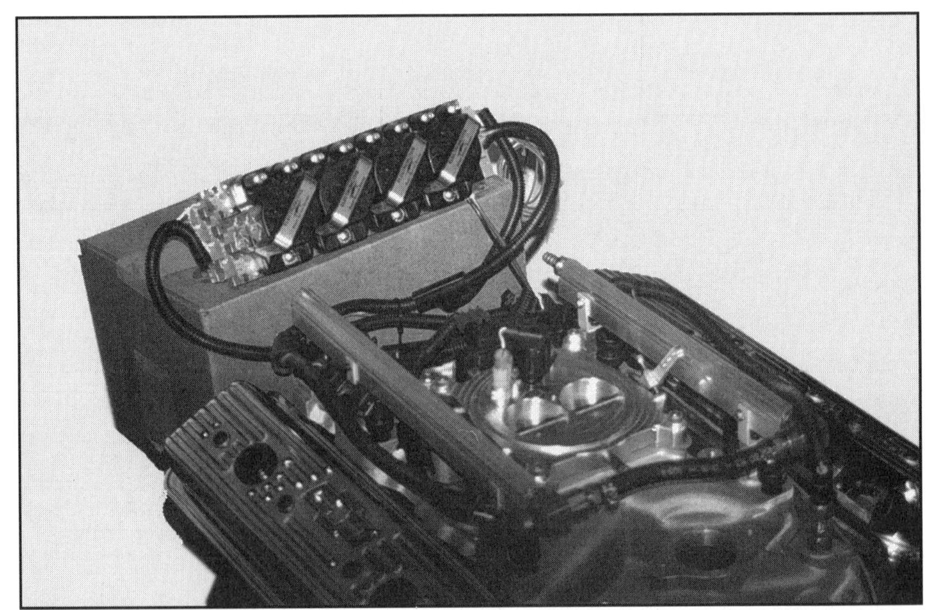

Electromotive kits can be fitted to any number of crate motor applications. This TBI throttle body installation illustrates the simplicity of the system. A laptop computer is required to make programming changes.

tors and fuel rail, closed loop EGO sensors, MAP sensor, air temperature sensor, and coolant temperature sensor. Throttle body, multiport and individual throttle injection packages are also available, and special software has also been developed for turbocharged and supercharged engines. The same basic systems are also incorporated on big block Chevys and 5.0L small block Ford engines. And because it is fully programmable, you can always recalibrate it for any modifications you may apply to the engine at a alter date.

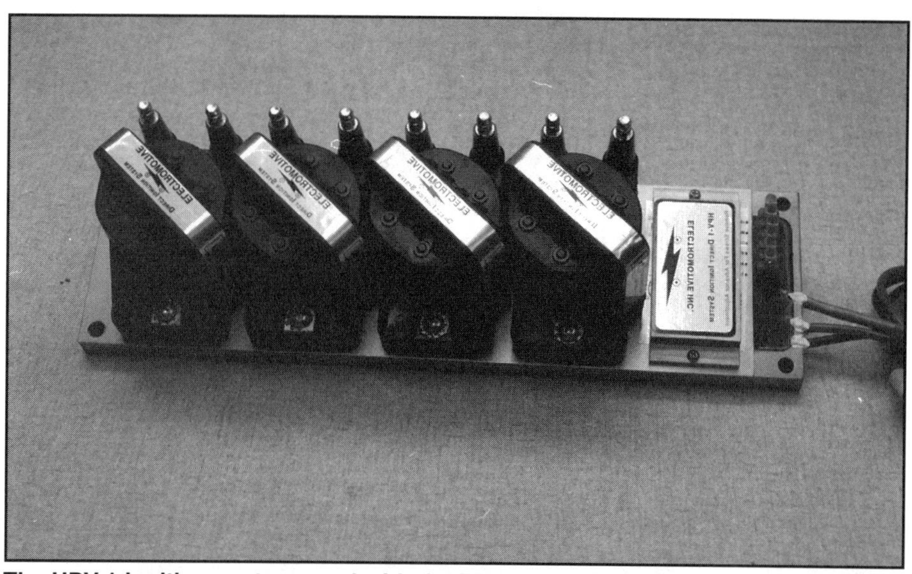

The HPV-1 ignition system used with the EFI setups provides fully programmable, crank triggered, direct ignition. Each coil fires two cylinders; one on the compression stroke and one on the exhaust stroke. Minimal energy is required on the exhaust stroke so no spark energy is wasted. This system, available on Electromotive crate motors provides dead accurate timing and full spark energy at all engine speeds.

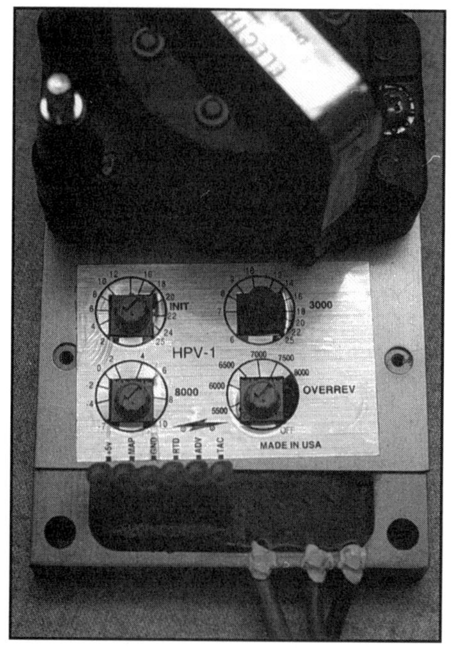

Spark timing adjustments are easy with the HPV-1 ignition system. The first knob (upper left) adjusts initial advance between 2° and 25°. The 2nd knob (upper right) adjusts 6° to 25° as the engine revs to 3000 rpm. The 8000 rpm knob (lower left) can add up to 10° additional timing or 7° of high speed retard. The last knob (lower right) sets the soft rev limiter between 5000 and 8000 rom, or it can turn off the rev limiter if desired. Expansion ports are also provided for tach output and or remote timing adjustments.

Electromotive systems use their own special trigger wheel to determine crankshaft position and speed information for the computer. Depending on the application, Electromotive can also provide a drop-in trigger assembly which replaces the distributor, or a special trigger assembly that mounts directly on the customer's old distributor.

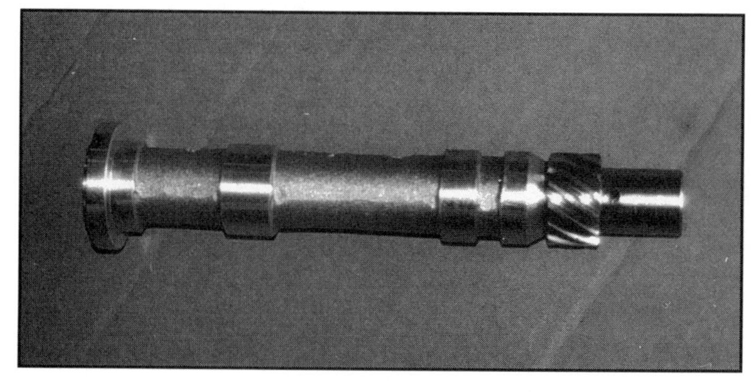

A stub distributor shaft is used to seal the original distributor opening on small block and big block Chevys. A distributor is no longer necessary with the distributorless ignition system.

SOURCE

Electromotive, Inc.
9131 Centreville Rd.
Manassas, VA 22110

(703) 378-2444
(703) 378-2448 fax

Crate Motor Buyer's Guide

Crate Motor Accessories

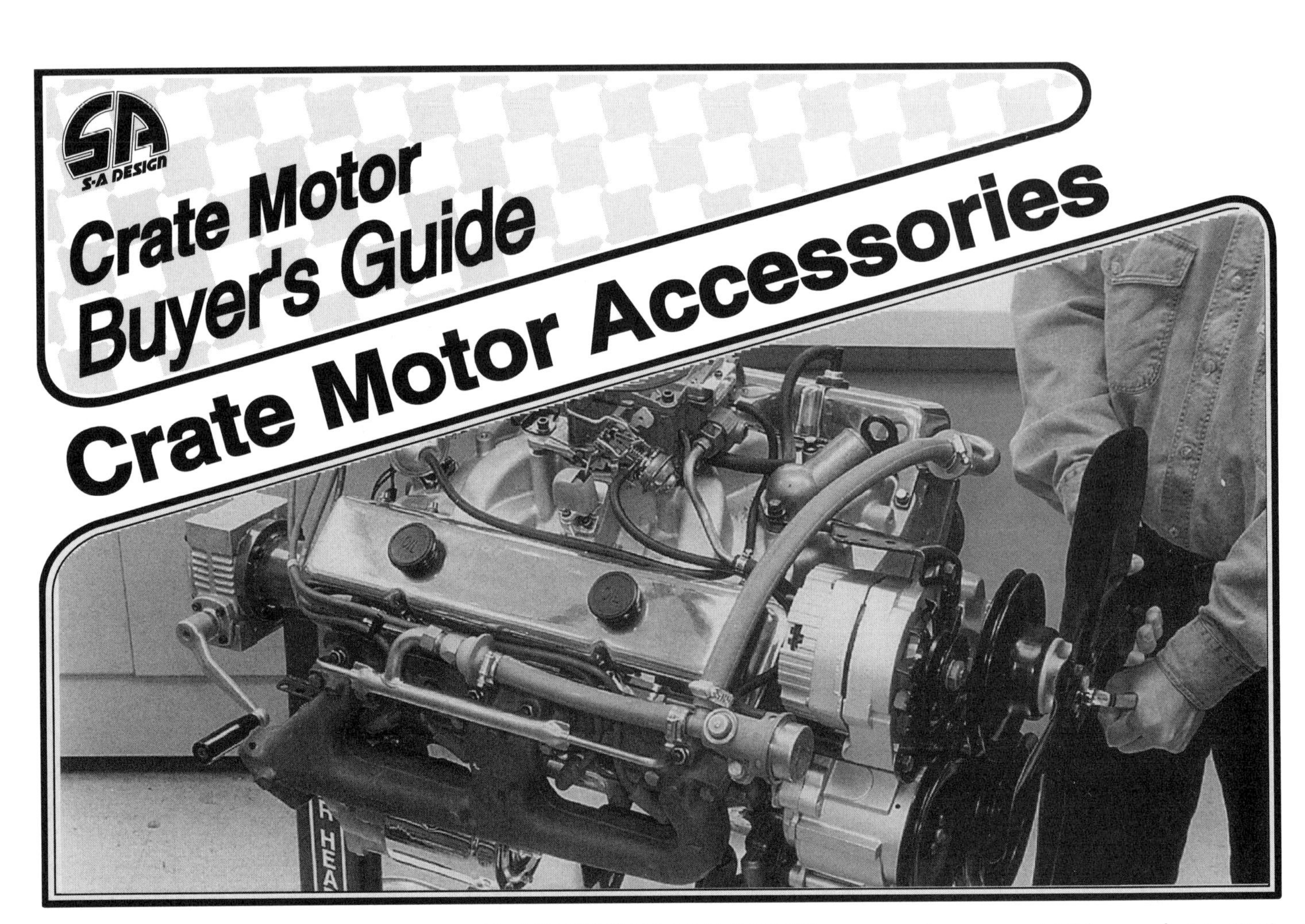

The biggest problem you may face when installing a new crate motor is the lack of proper accessories or other components that allow you to complete the job easily. Because crate motors are built by so many different engine builders, there is wide latitude in the way they are configured for the consumer. Some crate engines come with some accessories installed. GM motors for example may include a distributor, balancer and automatic transmission flexplate. Many after market motors also include these items, but just as many do not. When you make your purchase arrangements, make sure you understand just what you are getting and what you aren't getting with your crate motor purchase.

If you are purchasing a short block, an engine kit, or a long block assembly, you may require different components to complete the installation. In many cases the engine is being used to replace an existing engine that has failed or just plain worn out.

TRANSMISSION CONNECTION

Crate motor suppliers sell most of their engines for use in cars with automatic transmissions, hence most engines are offered with an automatic transmission flexplate installed. Not all engines are sold this way, but a great majority of them are so equipped. If you are running a manual transmission you need to make the salesman aware of it so he can specify the correct engine balance and supply a flywheel if you desire one.

The majority of crate motors available are shipped with an automatic transmission flexplate installed. When ordering, you must specify the type of transmission you intend to use and make certain you get the correct size flexplate or flywheel to accommodate your starter.

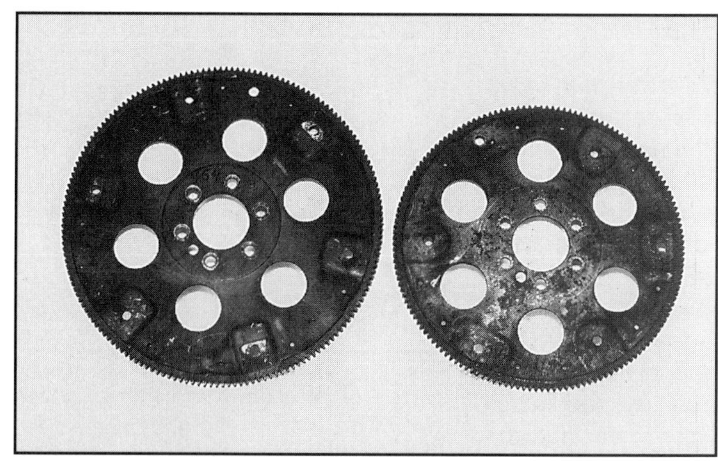

Flexplates and flywheels may be externally balanced or neutral balanced depending on the engine. There are also different diameters dictated by clutch or converter size and starter location. Make sure your salesman specifies the correct one.

Here is an externally balanced big block Chevy flywheel. Note the extra weight added at the upper left. If this flywheel were installed on a neutral or internally balanced engine it would cause a severe vibration.

The salesman also needs to know the type of transmission and torque converter if you run an automatic so he can match the flexplate properly with the converter and the starter. All the different diameter flywheels and flexplates can cause a problem if they aren't compatible with the rest of your parts. If you're planning to use an existing flywheel or flexplate, check it carefully for cracks or other damage. Make sure the flexplate isn't bent or otherwise distorted. If you are using a flywheel, now is the time to have it resurfaced if the clutch surface looks worn or damaged.

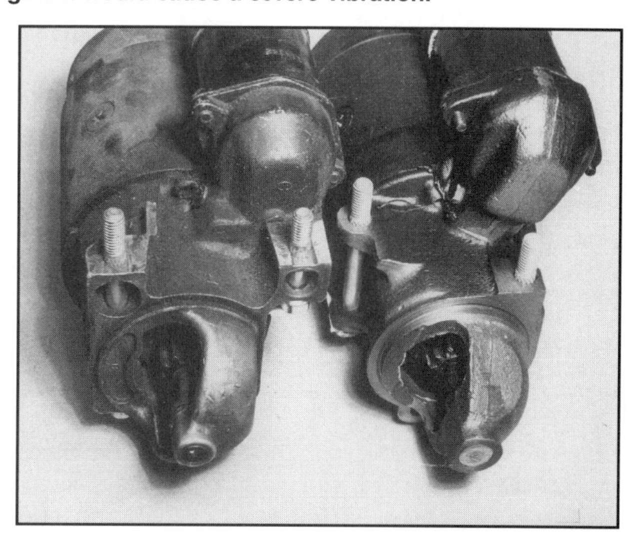

Starters come in al shapes and sizes including standard and heavy duty models. Some of them mount differently depending on the year of the engine. Some of them mount to the bellhousing instead of the engine. Your crate motor won't have a starter so you will have to transfer an existing starter or purchase the correct unit to complete your installation.

BALANCERS

If your new crate motor doesn't come with a harmonic balancer, you will have to supply one. Some engines are externally balanced with weighted balancers just like the flywheels and flexplates. Make sure you have to correct one. If you are going with a used balancer, check it carefully for damage or wear. Most balancers have a rubber lining holding the inner and outer hubs together. Make sure the rubber isn't cracked or otherwise deteriorated. The balancer could separate at high speed if this condition exists. Also check the inner surface where the balancer slides over the crankshaft snout. There is usually no damage, but you want to be certain that the positioning key is compatible with the crankshaft snout.

Harmonic balancers may also be internally or externally balanced. Externally balanced dampers are easy to spot because they have an obvious weight bias on one part of the balancer. Most crate motors come equipped with the correct balancer. If yours doesn't come with a balancer make sure you know how the engine is balanced so you can get the correct balancer. the salesman should be able to specify the correct balancer for you.

DIPSTICKS

It is pretty difficult for a crate motor builder to supply all the various types of dipsticks used in different vehicles. Many crate motors come with a universal dipstick designed to work with whatever custom oil pan the engine builder is using. If your crate motor doesn't come with a dipstick you'll have to determine whether the required dipstick and mounting tube are block mounted or pan mounted. The crate motor salesman may be able to recommend a dipstick and tube or you may have to purchase the correct factory combination to fit your specific chassis and engine combination at your local dealer. Most auto parts stores have universal dipsticks that may fit your applications with little or no modification. Dipstick incompatibility can be more of a problem than you might expect so be prepared to deal with it before you start your crate motor installation.

Some engines use a block mounted dipstick as shown here; other mount the dipstick tube on the oil pan. Either way, you have to have one to keep proper track of the engine oil.

BREATHERS

Almost all engines come with valve covers, but they may not have valve cover breathers. If the engine is a high performance street machine, aftermarket breathers are appropriate. Chrome and billet breathers are available through mail order, and the crate motor builder may even sell them separately. If you are looking at a late model replacement engine, breathers are not acceptable. You will need an oil filler cap on one valve cover and the correct rubber mount for a PCV valve on the other valve cover. Some engines don't mount the PCV valve on the valve cover so one cover may be left blank.

EXHAUST SYSTEMS

Outfitting your crate with a high performance exhaust system is the key to extracting maximum performance. Most crate motors respond very well to exhaust system modifications and you can definitely improve performance by complementing them with the right pieces. But that doesn't necessarily mean tube headers. There are many applications where a factory type exhaust system is much more desirable for a variety of reasons. The fact is that many engines don't really run all that badly with cast iron exhaust manifolds and in many cases these so called "junk parts" can work to your advantage. We've said it time and again, and it still holds true. You've got to have an integrated system, and if your application includes a requirement for quiet efficiency and/or the need to fit into a cramped engine compartment, then cast iron factory exhaust manifolds are just the ticket. They are still plentiful and the price is usually right.

The most obvious situation where cast iron manifolds are desirable is the street machine that is being built on a strict budget. This is particularly true when the new application involves an engine swap where there is not a specific set of headers available. In many of these cases an existing header can be slightly modified or rebuilt to fit the engine/chassis combination, but it's more likely that one of the many types of factory manifolds

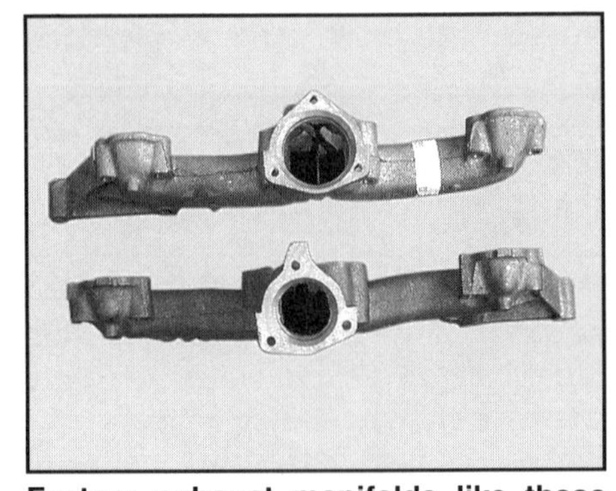

Factory exhaust manifolds like these early Corvette units are highly prized and very useful in many swap applications using crate motors.

will fit with very little difficulty. They are all designed to hug the side of the engine quite closely and they have either center exit or rear exit head pipe flanges.

For swappers who desire a clean, quiet installation, factory manifolds are the key. And in many cases performance can be boosted by using older factory exhaust manifolds that have larger openings. They are still available through many of the specialty restoration houses. Another aspect to consider is constantly pending legislation involving aftermarket performance parts. Someday in the near future it may be to your great advantage to use factory performance hardware to disguise your much modified small block from big brother's prying eyes. When and if that day comes, some of these pieces will suddenly become quite valuable.

HEADERS

Tube exhaust headers are a different story entirely and it's interesting to take a look at their development. Engineers have long understood that a major portion of an engine's energy passes out through the tailpipe in the form of heat and that the only effective way of utilizing this energy is through turbocharging, but they also knew that they could take advantage of the inertia of these spent gasses to help draw in the fresh intake charge. Much of the theory we discuss here is aided by the use of tube headers. The scavenging effect of a properly designed header can be a positive benefit to the induction system and to power output as a whole. The main thing to remember is that headers do much more than simply reduce back pressure across the engine. Of course they are less restrictive, but that doesn't always mean they make more power.

To obtain the absolute benefits of the so-called "scavenging effect," header designers discovered that the length, diameter and shape of both the primary pipes and the collectors has a great deal to do with the headers effectiveness. Since the primary tubes separate each of the engine's cylinders from the others, it offers the opportunity to let each cylinder help "scavenge" the others.

Typically there are two things happening inside a header when the engine is running. As each cylinder is exhausted, a pressure wave is created that travels down the length of the primary pipe at approximately 1700 feet per second. When the pressure wave reaches the end of the tube and exits into the collector, the sudden release of pressure creates a slight suction pulse that travels back up the header tube to the exhaust port where it actually helps to suck out any residual gasses and also draw in the incoming fuel charge for the next stroke. The suction pulse is further strengthened by the inertia of the exhaust gasses that are moving through the pipe at something in the neighborhood of 300 feet per second.

Naturally it is difficult to get all of these elements timed exactly right, and that is the chief function of making all the primary pipes equal in length. This is an important point that is much lauded, but seldom practiced. In most cases it is just too difficult or not financially feasible to make the primary pipes equal length. Some headers get really close and others miss by a mile without even apologizing. As it turns out the actual benefits are often difficult to discern unless you have access to a dyno or lots of track time for testing. In a race motor you have everything to gain because you have already built the motor with an eye toward equalizing power output in each cylinder. When this is the case a header system can work much more effectively. On the other hand, few street motors make the same power in each cylinder, so much of the scavenging effect is lost to the ill-timed suction pulses. That's why it is less important to have equal length tubes on a street header.

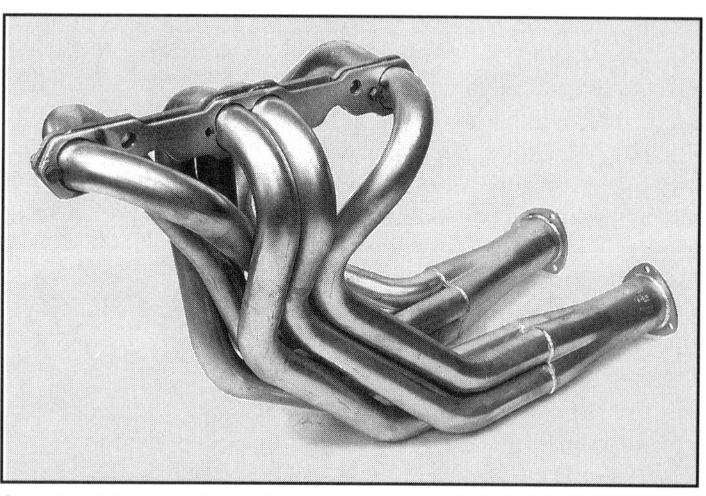

Standard "convenience headers" are widely available and they can provide good performance. Smaller engines should use 1-½-inch primaries while bigger engines need 1-⅝-to 1-¾-inch for small blocks and 2-inch for big blocks.

In fact it's fairly pointless to look for a street header that has equal length tubes, especially if you're going to route them through some sort of exhaust system anyway. (What we really find is that the reduction in back pressure will increase engine efficiency without specific regard to header pipe length.) Regardless of the claims, it would prove difficult for one manufacturer to actually demonstrate that his header design is far superior to another in a street-oriented application. It is true that a street headers design is important in terms of pipe diameter, collector length and collector diameter, but most of the manufacturers are pretty well dialed in on this stuff and the important differences center more around quality of construction, ease of installation and sparkplug access, and the useful life of the header.

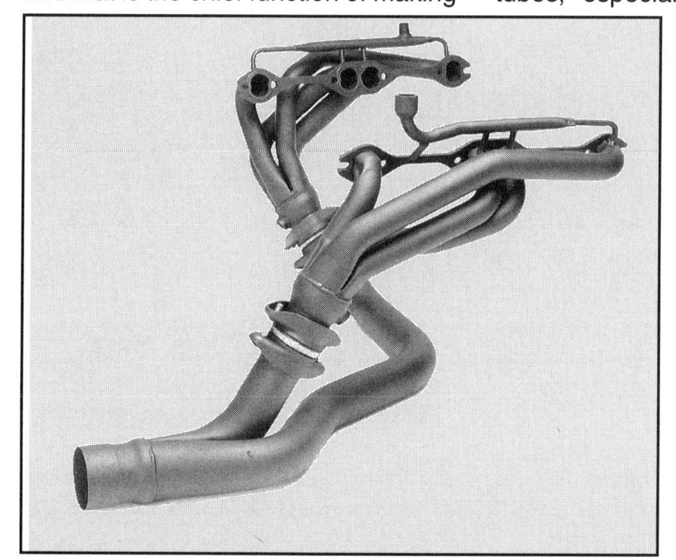

Tubular exhaust systems such as this setup by Edelbrock are just the ticket for late model swaps that need to maintain emissions compliance. Available coated and uncoated, they are compatible with existing catalytic converters.

This still doesn't mean that every header on the market is going to be a top performer in all respects. The major manufacturers all exercise pretty much the same approach. They know that a good street header is going to use small diameter primary pipes to help keep exhaust velocity high at slow engine speeds. The collectors will generally be longer than those on a race header since the longer collector helps to promote low-speed torque. The reasons for this are complicated, but it generally involves the headers ability to scavenge the cylinder. On a stock type motor it is often possible for the header to actually over-scavenge the cylinder and create a power loss by drawing some of the incoming fuel charge right though the cylinder without ever being burned. The longer collector acts something like a buffer to slow down the pressure wave velocity as it leaves the primary pipe, thereby allowing the engine to utilize fuel more efficiently. This is essentially accomplished by maintaining exhaust gas velocity, but decreasing the speed of the pressure wave.

Race motors follow a completely different tact. While primary pipe size has decreased on street headers, race headers continue to get larger; at least that is the case with top running Pro Stockers where the engines are putting out in excess of 700hp. When the engine is capable of making that much horsepower it can effectively use the 2-⅛-inch and 2-¼-inch headers that some of the Pros are running. But you have to remember that these are very high speed headers that don't really become effective until the engine is running well above 7000 rpm in a drag race configuration. Superspeedway cars are still running headers between 1-¾ inches and 2 inches in diameter even though they are starting to pull quite a bit of horsepower. They just can't use a larger header at their engine speeds.

It's plain fact that just about every bracket type engine will be perfectly happy with a 1-¾-inch-diameter header. Some of the really nasty ones may derive some benefit from a 1-⅞-inch header, and a 2-inch header is out of the question unless you're sure you're making over 550 hp at engine speeds above 8000 rpm, or running a big block. Don't get carried away.

If you don't use pre-coated headers—and you should—you can always protect your expensive headers and exhaust system components by spraying them with a high quality temperature resistant paint. For best results, apply several lights coats instead of one or two heavy coats that may sag or run and spoil the finish.

Make sure to use anti-seize compound on all exhaust system fasteners to prevent sticking and damaged threads. This is especially important where fasteners screw into aluminum, such as header bolts on aluminum heads.

DISADVANTAGES

Whatever your crate motor application, there are a number of disadvantages related to tube headers that you should consider when making your selection. There are also some tuning procedures that will make or break your installation so don't ignore them. The major problems with tube headers are not so much disadvantages as they are things you may wish to consider. They are of course noisier, but not much if the exhaust system is a quiet one. Many people feel that they run hotter, but what could be hotter than a cast iron manifold three hours after you shut the motor off. With headers you can grab the pipes after about half an hour. The chief problem with headers is usually leakage, breakage or rust or a combination of the three. These are all problems that can be controlled with a little maintenance and the exercise of a little common sense when shopping for your new pipes. In most cases a header that has a thicker port flange will last longer and leak less. Some makes offer thick-wall pipe for improved noise and corrosion resistance, and paramount in any street header selection is the ease of installation and access to spark plugs.

OTHER CONSIDERATIONS

When you install your headers, you should always paint them with Sperex VHT high temperature paint. This is the only stuff short of commercially available coatings that will stay in place and protect your headers from rust and corrosion. Many headers are already painted when they come out of the box, but don't let their shiny new appearance fool you. It is usually plain lacquer and it will burn right off as soon as you start the engine. The bolts are going to require retightening several times until the flange takes a set and the gasket settles down.

One of the biggest complaints about headers is that they don't seem to work after they are installed on an engine. This is usually because the installer neglected to retune the engine to complement the headers. Generally headers will require that the engine be jetted slightly richer and the improved scavenging might want a little less timing due to better combustion efficiency. It is often helpful to install a new set of spark plugs with the headers and run them for a few days before checking their color. Now you can't really tell much from plug color on a street engine since there are too many variables, but they will give a general indication of the engine's fuel mixture. If your old plugs were burning a medium tan color, the engine was probably running at close to the optimum air-fuel ratio. The header installation may lean the mixture enough to turn the plugs almost white. This indicates that you need to go anywhere from one to three steps richer with the jets. Normally you would increase jetting squarely, but for most applications you really need to re-jet the primary side of the carburetor only since that is the part of the carburetor you use most often in normal driving. As a rule you can jet the primaries first and then drive the car for some period of time to familiarize yourself with its requirements. If it wants richer secondaries you'll be able to tell after you get used to driving it.

THE COMPLETE SYSTEM

A fully-integrated exhaust system is going to continue beyond the end of the collector of the head pipe flange. It doesn't matter whether you're running cast iron factory manifolds or tube headers, performance will be improved if you keep back pressure in the exhaust system to a minimum. The very least you should do is equip the car with a dual exhaust system, but there is a right way and a wrong way to go about it. There are specific points to consider when constructing a truly effective dual exhaust system. The first thing you need is larger-than-stock exhaust pipes that are routed with a minimum of restrictive bends. Two-inch pipe is considered a good size, but if your engine is really healthy and you want to achieve maximum performance, you should use 2-¼-inch or 2-½-inch pipes. This is especially true if you are running a large displacement engine.

You'll want to augment the large diameter exhaust pipes with good free flowing mufflers. And if you really want the truth, this excludes side pipe mufflers and glass packs. They simply are not an effective alternative to a well-designed "reverse flow" muffler. The variety of available mufflers is considerable, but in reality there are a couple of commonly available units that really work, so why look for anything else? Most of the available "turbo" type mufflers will work very well, or you can go with the original turbo muffler used on the early turbocharged Corvairs. These units have full 2-½-inch inlets and outlets and they keep a car quiet while flowing very efficiently. Another excellent choice is AP's PN2592 which is basically a street hemi muffler. It is longer and it features 2-¼-inch inlets and outlets. Large diameter pipes and a set of these mufflers are really all you need, but there are a few other things you can do to improve the system's sound-absorbing capacity.

It is important that you use tailpipes behind the mufflers. If you have built an effective system, a tailpipe of the same diameter will not be a performance deterrent and it will aid immeasurably in keeping the car quieter. Another effective modification for noise reduction is the addition of a crossover pipe between the two sides of the dual exhaust system. This helps to equalize the pulses in each side of the system and make it quieter. The tube should be the same size as the exhaust pipes or, if space permits, it can be even larger for more effective dampening. It's location is not super critical, but you should strive to keep it within eighteen inches of the collector or head pipe flange.

Most complete crate motors will come with the valves already adjusted. Many have already been run-in or dyno tested to check the power. If your engine comes with a solid lifter camshaft it is a good idea to at least check the valve lash prior to the initial startup.

Crate motors rarely come with fuel pumps, but most will come with the mounting plate in place for a standard mechanical pump. you will have to supply all the fuel lines and connections for the fuel system.

Crate motors are frequently upgraded with bolt-on power kits such as a supercharger or nitrous oxide injection. When you do this, you need to be sure that the engine's internal components are up to the extra stress of increased cylinder pressure.

Prepping A Crate Motor

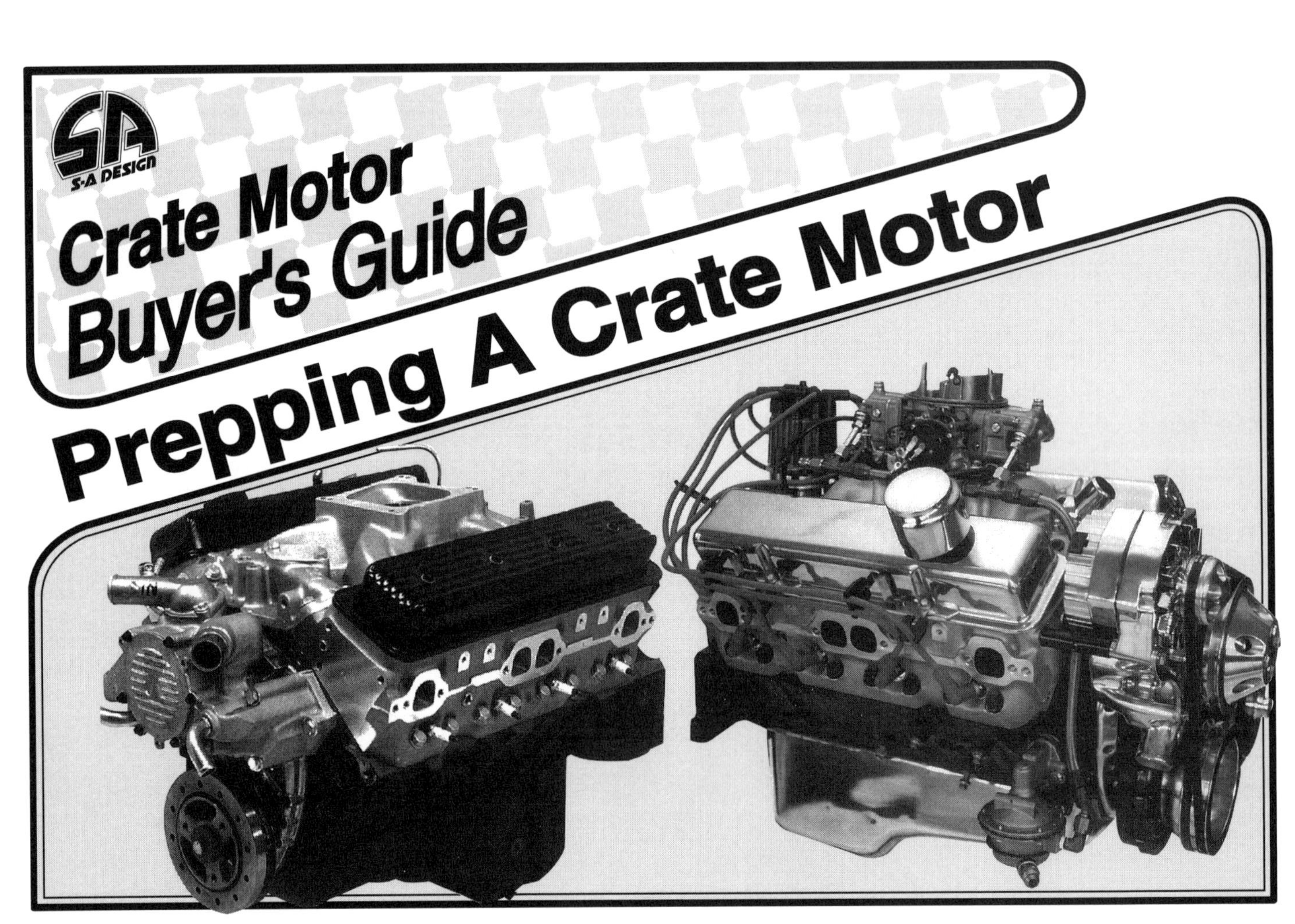

Installing your new crate motor can be a fairly trouble free experience or your worst nightmare depending on the application. Even though you may have purchased a fully assembled engine, there are still numerous checks, adjustments and pre-startup considerations necessary to ensure a clean start and a proper break-in.

Most crate motors are purchased as complete engines less carburetor, distributor, starter, water pump, motor mounts and so on. In the overwhelming majority of cases, the buyer is able to simply transfer all the appropriate accessories from his previous engine. If the engine is purchased as a long block assembly it will typically have cylinder heads and valve train, but no intake manifold. A short block assembly will require selection and installation of cylinder heads.

Whatever the case, you need to check your new engine out thoroughly to make certain it is undamaged and that all the proper adjustments have been made. While it probably isn't necessary, you may want to check the cam timing to make sure the cam is in the correct position. With the engine mounted on an engine stand or even in the shipping crate you can attach a degree wheel to the harmonic balancer or the crank snout if no balancer is included. A simple screw-in piston stop in the number one spark plug hole will allow you to establish top dead center and adjust the degree wheel to verify the cam's opening and closing points and the intake centerline (see accompanying sidebar on cam basics).

HOW TO DEGREE A CAM

Degreeing the cam ensures that it is installed in the correct position relative to the crankshaft and gives you a baseline from which to determine valve-to-piston clearance. Although rather rare, out-of-the-box cam phasing can be off for any number of reasons, including an improperly placed dowel pin or a misplaced keyhole in the cam gear. This can affect performance and reliability unless it is observed and corrected before the engine is started.

There are several ways to degree a cam, but the simplest method may be the one we describe here from Competition Cams called the "Lobe Center Line" method (Note: this method works only with cams that have symmetric lobes; asymmetric profile cams must be degreed using the cam grinder's 0.050-inch lift figures). Regardless of method you

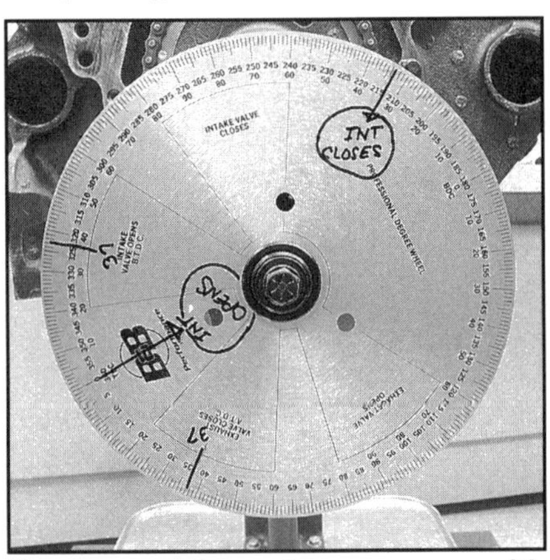

If you are degreeing a cam with asymmetric cam lobes, you cannot use the "Lobe Center LIne" method. After precisely setting the degree wheel to TDC as indicated in the text, mark the 0.050-inch timing figures on your degree wheel and verify them with your cam card.

select, the first step is always to accurately locate TDC on the number-one piston (front piston on the left—driver's side—bank). The quickest method is to affix a piston stop over number-one cylinder, then adjust and lock the stop bolt so the piston will be stopped about 1/2-inch from the top of the bore. Attach a degree wheel to the snout of the crank, and firmly fasten a wire pointer to the block so that it points to the degree marks on the wheel. Start off by positioning and locking the degree wheel with pointer on BDC when number-seven piston appears to be at top dead center (just eyeball it for now). Then rotate the engine until number-one piston contacts the stop; mark the reading on the degree wheel). Then, rotate the engine in the opposite direction, and when the piston once again contacts the stop, note this second reading on the wheel. Now that you have a reading on either side of TDC, add them together and divide by two (finds the average). For example, if you read eighteen degrees on one side and twenty-one degrees on the other side, the average would be 19-1/2 degrees. This means that true TDC is located exactly 19-1/2 degrees from both of the marks on the degree wheel. Remove the piston stop and rotate the engine until pointer indicates 19-1/2 degrees between each of your marks and the pointer. Double check your observations, then rotate the degree wheel (not the crankshaft) so that the pointer indicates exactly TDC, then lock everything down tight.

Now that TDC has been determined, we'll proceed with the Lobe-Centerline Method to degree-in the cam (remember, this method only works with symmetrical lobe profiles). First, we'll locate the point of maximum lift on the intake lobe. Place a dial indicator on the number-one intake lifter and rotate the engine until the maximum-lift point is indicated and zero your indicator dial. This will only be an approximation, and like finding TDC, you'll have to sneak up on it. Continue rotating the engine slowly until the lifter once again begins to climb the lobe. This time stop when the indicator reaches 0.050-inch before max. lift and take your first degree wheel reading. Rotate the engine further and when the indicator reaches 0.050-inch after observed max.-lift point take a second degree wheel reading. The point exactly halfway between these readings is the true max.-lift point and it is also the intake lobe centerline.

Now all you have to do is observe where the centerline resides in relation to TDC. Count the number of degrees from this point to TDC and compare it to the number on the timing card that came with the cam. If it indicates that the cam should be installed at 110 degrees, the centerline should be 110 degrees from TDC. If you find that it is only 108 degrees from TDC, the cam is retarded and you will need a two-degree bushing in the cam gear to advance it to the 110 degree position. This can also be accomplished with an offset crank key or one of several other methods.

With the cam degreed as designed by the manufacturer, you can check other critical valve train clearances, including valve-to-piston clearance.

CRATE MOTOR Performance Tips & Tricks

CAMSHAFT BASICS

Camshaft terminology can be confusing, so here's the low-down. To start off, the camshaft is a round shaft incorporating **cam lobes**. The **base circle diameter** is the smallest diameter of the cam lobe and is shaped perfectly round. **Clearance ramps** begin the transition from the round base circle to the **flanks** of the lobe. As the cam turns, the lifter is smoothly accelerated by the clearance ramp and flank, continues to rise as it approaches the **nose**, then begins to slow to a stop as it reaches maximum lift at the **lobe centerline**. Maximum **lifter rise** is determined by the height of the tip of the cam lobe from the base circle diameter. The lifter then accelerates in the closing direction and when the valve approaches its seat, the lifter is slowed down by the flank and closing clearance ramp. Valve duration is the number of crankshaft degrees that the lifter is held above a specified height by the cam lobe (usually 0.006-, 0.020, or 0.050-inch). A **symmetric lobe** has the same lift curve on both the opening and closing sides; an asymmetric lobe is shaped differently on each side of the lobe. A **single-pattern cam** has the same profile on both the intake and exhaust lobes; a **dual-pattern cam** has different profiles for the intake and exhaust lobes.

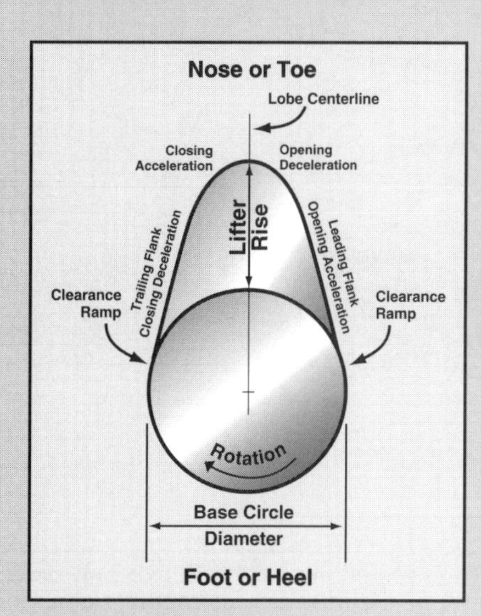

Lobe-separation angle is the angle measured in camshaft degrees (multiply by two for equivalent crankshaft degrees) between the maximum-lift points on the intake and exhaust lobes for the same cylinder. The lobe-center angle is "ground" into the cam when it is manufactured and cannot be changed (unless the cam is reground). As the lobe-center angle is decreased, the valve overlap period (when both intake and exhaust valves are open) is increased. This usually results in a rougher idle, reduced low-speed power, and increased high-speed power. The lobe centerline is the angle measured in crankshaft degrees between the point of maximum lift on the number one intake lobe (usually) and Top Dead Center. This value is determined by the "indexing" of the cam to the crankshaft. A reduced centerline angle (less than the lobe-center angle) indicates that the cam is advanced.

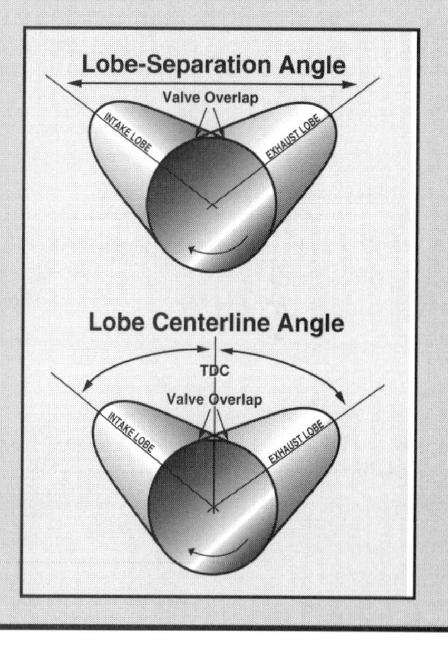

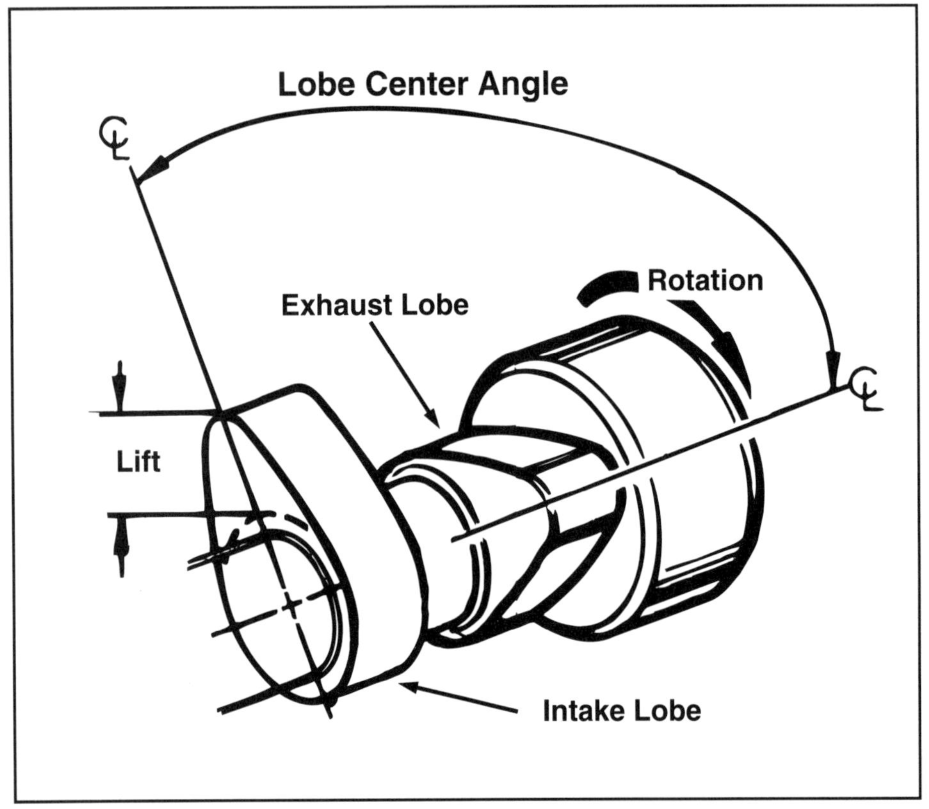

Lobe-center angle is measured between the centerlines of the intake and corresponding exhaust lobes. It is fixed when the cam is ground and cannot be changed (without regrinding the cam). A larger angle produces less valve overlap and boosts low-speed torque. Decreasing the angle yields greater valve overlap and moves the torque curve higher in the rpm range.

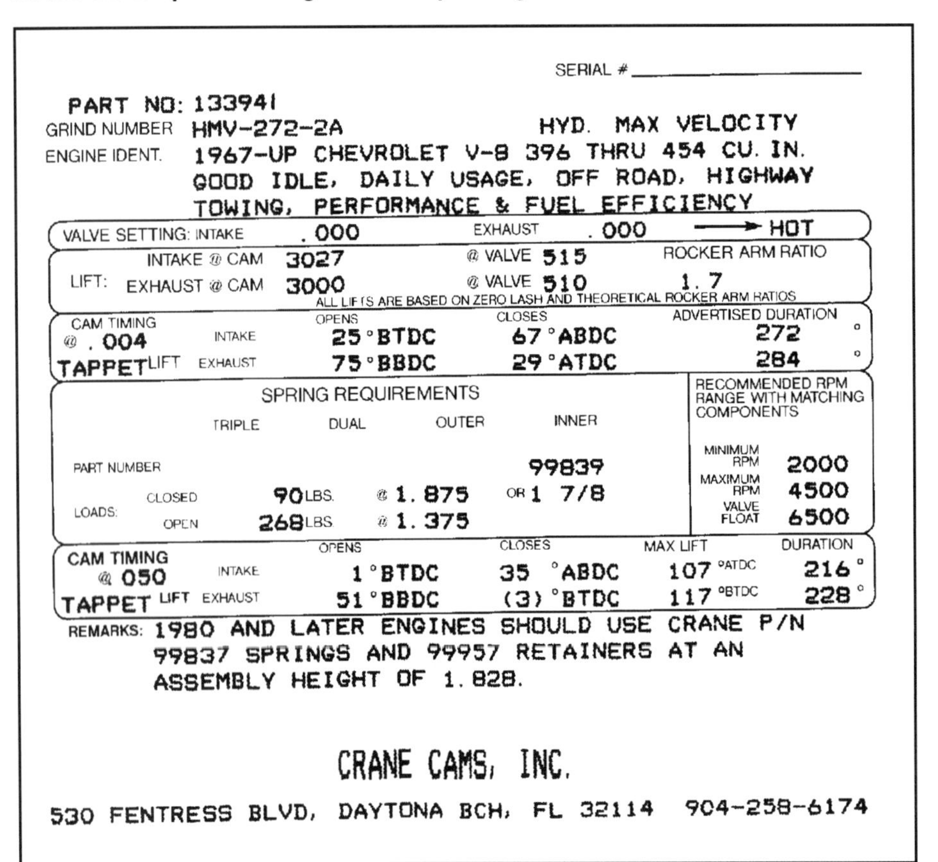

SERIAL # __________

PART NO: 133941
GRIND NUMBER HMV-272-2A HYD. MAX VELOCITY
ENGINE IDENT. 1967-UP CHEVROLET V-8 396 THRU 454 CU. IN.
GOOD IDLE, DAILY USAGE, OFF ROAD, HIGHWAY
TOWING, PERFORMANCE & FUEL EFFICIENCY

VALVE SETTING: INTAKE .000 EXHAUST .000 → HOT

LIFT: INTAKE @ CAM 3027 @ VALVE 515 ROCKER ARM RATIO
EXHAUST @ CAM 3000 @ VALVE 510 1.7
ALL LIFTS ARE BASED ON ZERO LASH AND THEORETICAL ROCKER ARM RATIOS

CAM TIMING @ .004 TAPPET LIFT	OPENS	CLOSES	ADVERTISED DURATION
INTAKE	25°BTDC	67°ABDC	272°
EXHAUST	75°BBDC	29°ATDC	284°

SPRING REQUIREMENTS

	TRIPLE	DUAL	OUTER	INNER
PART NUMBER				99839
LOADS: CLOSED		90 LBS.	@ 1.875	OR 1 7/8
LOADS: OPEN		268 LBS.	@ 1.375	

RECOMMENDED RPM RANGE WITH MATCHING COMPONENTS
MINIMUM RPM 2000
MAXIMUM RPM 4500
VALVE FLOAT 6500

CAM TIMING @ 050 TAPPET LIFT	OPENS	CLOSES	MAX LIFT	DURATION
INTAKE	1°BTDC	35 °ABDC	107 °ATDC	216°
EXHAUST	51°BBDC	(3) °BTDC	117 °BTDC	228°

REMARKS: 1980 AND LATER ENGINES SHOULD USE CRANE P/N 99837 SPRINGS AND 99957 RETAINERS AT AN ASSEMBLY HEIGHT OF 1.828.

CRANE CAMS, INC.

530 FENTRESS BLVD, DAYTONA BCH, FL 32114 904-258-6174

Your crate motor may or may not come with a cam card. If it doesn't have one, the supplier may be able to fax you a copy of it if you call them. You must have it to degree the cam properly. It contains a great deal of information about the cam.

Keep in mind that if you alter cam timing or even valve lash, it can have a substantial effect on valve-to-piston clearance, usually within ten degrees of TDC, particularly if you have an engine with tight valve-to-piston clearances. When you advance the cam, it will reduce valve-to-piston clearance on the intake valves. Retarding the cam moves the exhaust valves closer to the pistons.

Next to the cylinder heads, the camshaft and valve train represent one of the most important controls you have over engine output. There are remarkable power gains to be had from the right cam profile, if it is well matched to other engine components. By combining the correct cam timing with well integrated intake and exhaust systems, it is possible to achieve truly startling torque and horsepower. But remember, nothing does a better job of fouling up your engine's performance potential as a mismatched camshaft.

LOBE CENTERS

Imagine a line passing from the center of the cam directly through the highest point on the lobe. This is the geometric centerline of that particular lobe. To avoid confusion when comparing cams, remember that lobe-center angle is measured between the centerlines of the intake and corresponding exhaust lobes, while lobe centerline is the angle measured between the center of the lobe and TDC. Lobe-center angle is fixed and cannot be changed once the cam is ground. Lobe centerline, however, can be changed by advancing or retarding the cam. When you do this you are effectively moving the intake lobe centerline closer to or farther from top dead center TDC.

In terms of engine performance, the lobe-center angle is significant. A larger angle yields less valve overlap (the period when both valves are open). This permits the cylinders to begin building pressure sooner and that boosts low-speed torque. Decreasing the angle yields greater valve overlap and moves the torque curve higher in the rpm range. The effective range of rpm where the engine is an efficient power producer is also narrowed.

For most street applications always select a cam that will build as much torque as possible. Generally you want valve events that produce a wider lobe-center angle, decreasing valve overlap. One of the advantages of the new high-velocity factory and aftermarket street roller profiles is that you have good idle quality by using wider 112° to 115° lobe centerlines but give high-speed power a boost with a more aggressive lobe profiles (increased effective duration). The result is a broad torque curve, ideal for street use.

Turbocharged or supercharged engines should avoid cams with too much overlap since the pressurized intake system already provides effective cylinder filling and forced exhaust scavenging. In this application long overlap can be detrimental since some of the intake charge can be blown right through the engine without being burned.

For the average street and strip enthusiast all of these factors are taken care of by the cam manufacturer. Their vast experience lets them provide you with a cam that they know will work. The worst thing you can do is let your friends pick your cam. The best thing you can do is call the manufacturer and talk with their technical assistance rep. The price of the call is more than worth the help you'll receive. If the tech rep is friendly, sincere, and asks a lot of questions, chances are he really has your interests at heart and will supply you with a cam just right for your needs.

While you're discussing cam specs with the tech man, you might keep some other factors in mind. Depending on the application, you'll probably be using either hydraulic or solid lifters. For a variety of reasons, roller cams are an excellent choice, but most people still rely on flat-tappet grinds. Camshaft companies now offer great street roller profiles with either hydraulic or solid roller lifters that offer more performance than virtually any streetable flat-tappet cam.

CRATE MOTOR Performance Tips & Tricks

CRATE MOTOR HEAD GASKET TECH

When selecting gaskets for your short block crate motor, make sure the gasket bore is large enough so that it does not protrude in the cylinder. If it does, it will be erroded by combustion and fail in short order.

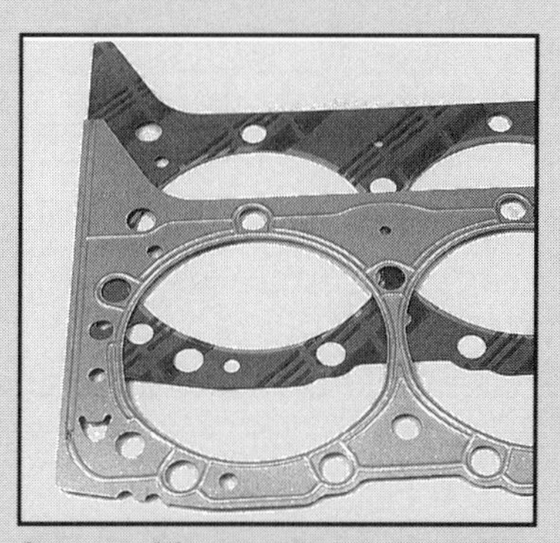

Composition gaskets are made of layers of material (back) and are suitable for both iron and aluminum heads. Steel-shim gaskets (front) have raised sealing beads around the critical openings and should only be used on cast iron heads.

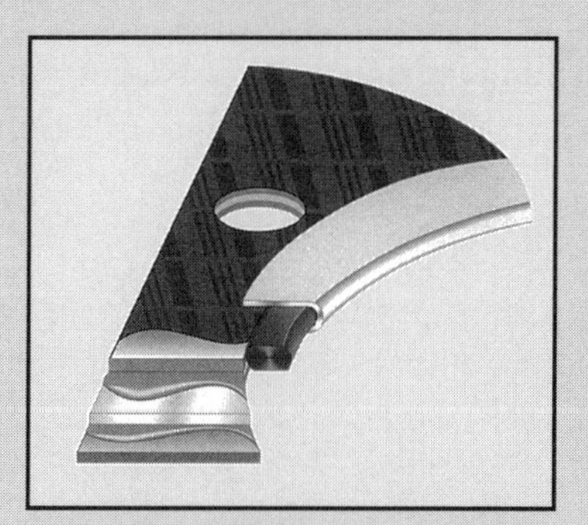

The latest Fel Pro composition gaskets for aluminum heads on high-performance engines have pre-flattened steel wire "fire rings" to optimize combustion pressure sealing without damaging the head surface.

Gasket Basics: Before you select a head gasket for a short block, make certain that it fits the cylinder bore (no overhang within the cylinder). Avoid universal gaskets that do not precisely fit the bore. Since gasket thicknesses vary, and the piston-to-head clearance is critical, you must know your engine's requirements before you can make a selection. Minimum clearance is 0.035-inch for engines with steel connecting rods. Pick a gasket that will give close to the minimum clearance, since this will usually optimize combustion chamber quench and flame travel.

Gasket Material: Composition head gaskets (made of layers of material) are suitable for both iron and aluminum heads while steel-shim (single sheet of steel) gaskets have raised sealing beads around the critical openings and should only be used on cast iron heads. Shim and composition type head gaskets are available from factory and several aftermarket sources such as Fel Pro, McCord, Victor, and Detroit Gasket. The best composition gaskets for aluminum heads on high-performance engines have pre-flattened steel wire "fire rings" to optimize combustion pressure sealing while minimizing brinneling (marking the aluminum head surface). The best steel-shim gaskets for iron heads are pre-coated with a silicone-like material to prevent water leaks.

Recommendations: Fel Pro gaskets are manufactured for all popular crate motor combinations and are frequently used in crate motor buildups. Fel Pro offers all the special gasket combinations that have been developed for special high performance factory and aftermarket applications.

Many production 4.000-inch bore Chevy engines use steel-shim gaskets. These 0.026-inch thick gaskets are available as GM PN 3830711. The composition head gasket for the same bore size is GM PN 10105117 and is 0.028-inch thick. This gasket is recommended for stock and moderately modified street and marine engines. GM PN 14088948 is a 0.051-inch thick composition gasket for the Corvette and HO aluminum cylinder heads. It can be used on both iron and aluminum heads. A 0.039-inch composition-style gasket with a stainless steel jacketing is available for Chevy smallblocks as GM PN 10159455. This is a good choice for high-performance street/strip combinations. Heavy-duty competition gaskets from Fel Pro are a Teflon-coated, composition design with solid "fire rings." Available under GM PN 10185054, they are 0.040-inch thick and fit 4.000- to 4.125-inch bores.

CARB SELECTION

When using any factory production manifold or a low-speed specialty manifold every effort should be made to keep carburetor size to a minimum. Single-plane manifolds, in particular, rely on increased mixture velocity to improve efficiency and large carburetors wholly defeat their purpose. When these manifolds are used on any engine, no matter what the displacement, carburetor sizes should be kept under 750cfm. In most cases a 600cfm carburetor will substantially improve the engine response and overall performance. For smaller engines in the 350 cubic inch range, a Quadrajet will really shine on the street and many people are having great success using the 750cfm 3310 Holley four-barrel with vacuum secondaries on these engines. This is particularly true of cars with automatic transmissions. In these instances the small carburetor provides crisp throttle response and remarkably brisk acceleration.

In many applications you should take care not to overlook the advantages offered by retaining the stock

The Quadrajet is back! Edelbrock offers four new Quadrajets in 750 and 800cfm versions. Manufactured for Edelbrock by Weber, they reflect a very high degree of quality and make an excellent performance carburetor right out of the box. Manual, electric, and hot-air choke models are available.The Quadrajet is back!

CRATE MOTOR Performance Tips & Tricks

SELECTING HOLLEY CARBURETORS

Holley four-barrels are one of the most versatile carburetors available for high performance work. They are available in a broad range of CFM ratings and Holley offers special configurations for street, high performance and racing applications. A Holley four barrel on top of your engine is still a universal sign that you mean business. This 990 CFM version was prepared by the Carburetor Shop in Rancho Cucamonga, California.

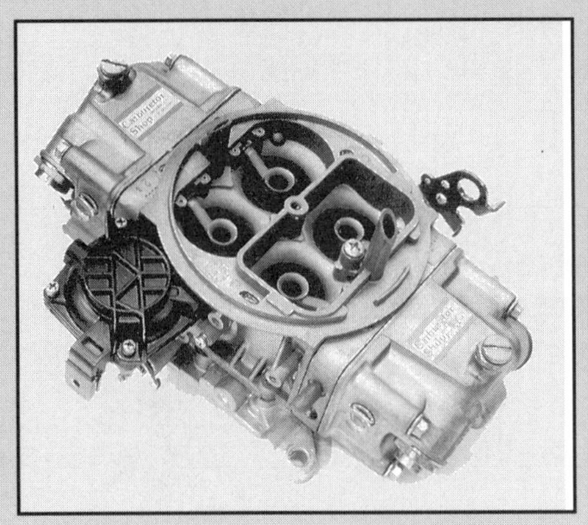

Holley's List 3310 vacuum secondary four-barrel is by far the most desirable Holley carburetor for street performance, RV applications and some circle-track racing. The 3310 has vacuum secondaries that allow a crisp transition from a well tuned primary circuit. It can be tailored to suit the unique requirements of nearly any application. If you want good drivability and throttle response in a street carburetor, the 780 CFM 3310 is your best choice.

For racing applications and "outer limits" street machines, the Holley Double Pumper is the leading contender. Available in sizes ranging from 600 CFM to 850 CFM, Double Pumpers feature an accelerator pump on both the primary and the secondary metering circuits. They can be configured with adjustable 4-corner idle circuits for precise tuning, and they are easily modified to suit any gasoline or alcohol racing application.

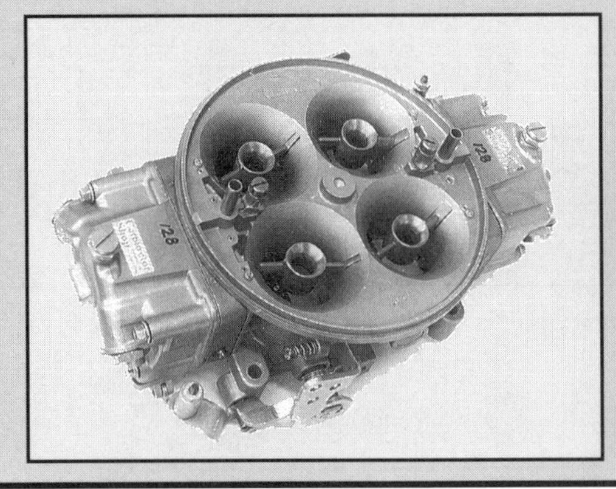

Holley Dominators are pure racing carburetors designed to flow massive amounts of air. Most of them feature an intermediate idle circuit in addition to the two standard metering circuits and they are quite tunable. The Carburetor Shop builds them in a range of sizes all the way up to 1425 CFM. The 750 CFM Street Dominators can be modified for strong street performance or racing.

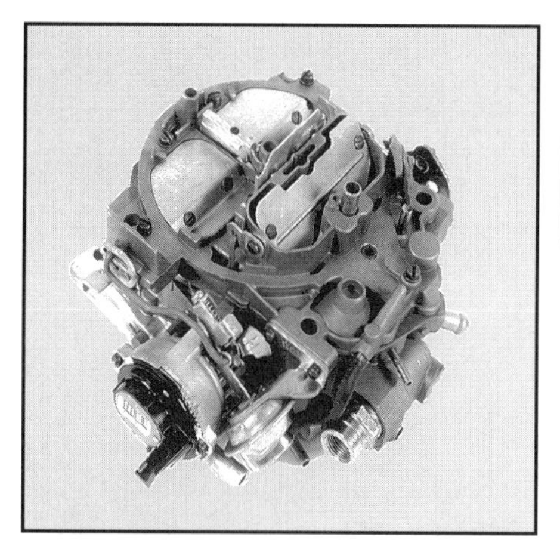

The Carburetor Shop offers new and remanufactured Quadrajets for all popular models. They also rebuild thousands of Quadrajets every year. They repair, upgrade, and restore Quadrajets for a reasonable fee. Their optional ColoRestore brings back the original factory color.

The basic differences between a single and dual-plane manifolds are clearly illustrated here. The dual-plane (top) divides the plenum in half, with runners grouped alternately by firing order. Each cylinder "sees" only one-half of the carburetor, transferring a strong signal to the boosters for good low- and mid-range power. The single-plane (lower photo) has short, direct runners that give access to the entire carburetor. This design is better at delivering consistent mixture distribution and is a consistent winner for high-rpm power. If you purchase a fully assembled engine it will come with the appropriate intake for the advertised power rating. If you purchase a long block or a short block, you'll need to consider the application carefully in order to determine the need for either a dualplane or a single plane intake manifold.

CRATE MOTOR Performance Tips & Tricks

FUEL SYSTEM REQUIREMENTS

High performance fuel systems are essential for any application with increased fuel demands. The old tried and true Holley Blue pump and regulator are still a top choice for basic bracket racing applications. The Carburetor Shop's Super-Blue version or Holley's Volu-Max are the next logical step up for increasing fuel system demands.

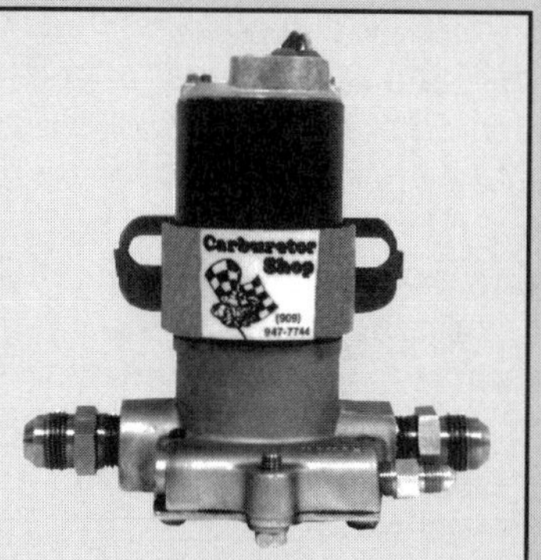

Maximum performance, high demand applications call for a Magna-Flow SP4450 pump with integral regulator and adjustable internal bypass, or the SP4401 pump without filter. Both pumps have high performance windings that won't overheat due to continuous running. Mechanical pump applications can use the modified 3 valve Carter pump from the Carburetor Shop. It delivers about 115 gallons per hour and a higher flowing 6-valve Clay Smith mechanical pump delivering over 130 gallons per hour is also available from the Carburetor Shop.

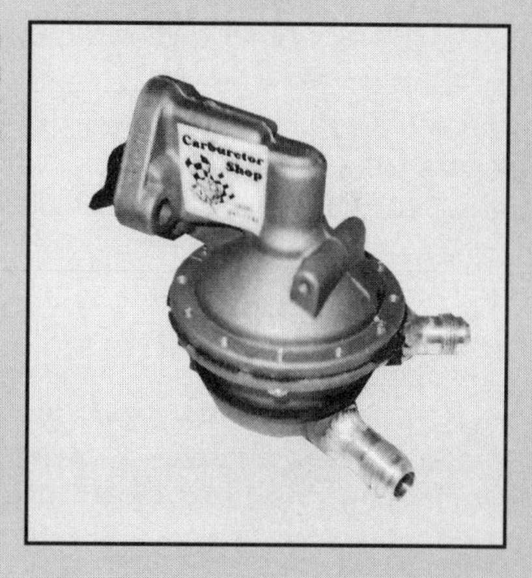

The standard Holley 2-port regulator is adequate for many high performance applications, especially in its modified form as found in the Carburetor Shop Super Blue pump kit. For higher performance applications or multiple carb setups the Holley 4-port regulator will get the job done. These regulators are good basic fuel regulators that do the job just as well as the multicolored regulators that cost two and three times as much.

carburetor. A new or remanufactured Quadrajet from Edelbrock or the Carburetor Shop often eliminates an expensive carburetor change. The Q-jet was offered as standard equipment on many Detroit medium-performance engines and it makes a good performance carburetor, especially in applications where low-speed efficiency and throttle response are an important consideration.

The key consideration is venturi diameter. You will note that carburetors with smaller maximum flow ratings have small primary venturis, producing increased mixture velocity and throttle response. If you're going to use a high performance carburetor, you may wish to use venturi diameter as a basis for comparison. Once again, the goal is to optimize carburetor selection for your individual car/engine combination.

Bear in mind right from the start that a carburetor that is too large will hurt performance far more than one that is too small. If you use a carburetor that is too small, you will be able to feel the engine "go away" when the rpm surpasses the carburetor delivery capacity. The engine will still run fine, but it will feel flat at the very top end. On the other hand, a carburetor that is too large will not send an effective signal to the metering circuits, reducing overall performance and fuel economy. If it is greatly oversized, the low end and low-speed acceleration will practically disappear. Severe cases even suffer from reduced power in the mid-range. This is far more difficult to detect since the engine may still feel "responsive." It just doesn't pull as hard as it should when the car is accelerating.

CRATE MOTOR Performance Tips & Tricks

MANIFOLD INSTALLATION TIPS

Before installing a manifold on a new short block, set the manifold on the block with the runner gaskets in place. Make sure the bolt holes line up; there is no shortage of mis-machined manifolds! Also make sure there is 0.060- to 0.120-inch clearance between the manifold and block rails (arrows).

Numerous factory and aftermarket gaskets have been devised to seal the gap between the intake manifold and the cylinder block at the front and rear of the engine. No method is as effective as a simple 1/4-inch bead of high temperature RTV silicone sealer laid across the end rails. Add a little extra around the water openings for extra insurance against water leaks.

To ensure an effective seal, be careful not to move the manifold around as it contacts the silicone. When you tighten the bolts the silicone squeezes out from the gap, forming a seal that will virtually never leak oil. Install the intake manifold bolts and torque in 5 pound-feet steps to 25- to 35 pound-feet. Follow the tightening sequence shown on page 89.

FINE TUNING CARBS

Carburetors nearly always work well in out-of-the-box form, but their efficiency can usually be improved with minor jet changes, idle adjustments, power-valve alterations, and other adjustments that can dial-in the combination. This fine tuning process must be performed with careful consideration of other engine and drivetrain components, e.g., gearing, transmission type, tire diameter, camshaft profile and the like. Careful evaluation is the name of the game. The engine will reveal what it does and doesn't like, but it is up to you to recognize and interpret the signals.

It's beyond the scope of this publication to delve deeply into the science of carburetor tuning. (The books Selecting Holley Carburetors, Rebuilding Holley Carburetors, Carter Carburetors, and How To Build Horsepower available from S-A Design cover this topic—see page 144 for ordering information.) You can, however, generally depend on factory calibration to be pretty close for most day-to-day applications. If you want to experiment some, set your sights on good off-idle response and crisp reaction in the mid-range, but don't overcompensate for problems that may be related to improper gearing or a lazy spark advance.

On all reasonable street applications you will have better luck with a carburetor that utilizes vacuum secondaries or some sort of vacuum air-valve and secondary-lockout system. The popular Holley double-pumper carburetors are designed strictly for racing. They work most effectively in relatively light cars with very low gears and in racing conditions where the carb is constantly transitioning into the secondary throttles. In a true racing setup the double-pumper is nearly impossible to beat, but for the street, stick with vacuum secondaries. A careful workbench mechanic can tune a vacuum carb for outstanding day-to-day performance but you must write down every tuning change and what specific (hopefully, measured) effect it has on performance. If you're in doubt about any change, make a careful back-to-back test to see if the alteration actually improves performance and/or economy. Be methodical and take your time.

CHOOSING AN IGNITION

Selecting an ignition system for your crate motor requires the consideration of several important factors. Since timing is the basis of all engine functions it is critical that, above all else, the ignition system maintain rock-solid integrity. This was easy enough in the early days when the primary mission of the ignition system was providing smooth engine operation, but now that it has been called upon to help control emissions and to compensate for other less-than-ideal conditions, the standard ignition system has become quite complex.

Potential horsepower increases from late model, high energy ignition systems whether factory or aftermarket depends on the flame propagation characteristics of the combustion chambers. Cylinder heads with larger chamber volumes can benefit most from these ignitions, and although it's impossible to predict the benefits on any single engine, gains may vary from negligible to as much as 5%. Engines with small-volume chambers usually show little or no improvement from multi-firing ignitions. However, multi-

CRATE MOTOR Performance Tips & Tricks

BOLTS, STUDS & TORQUING TIPS

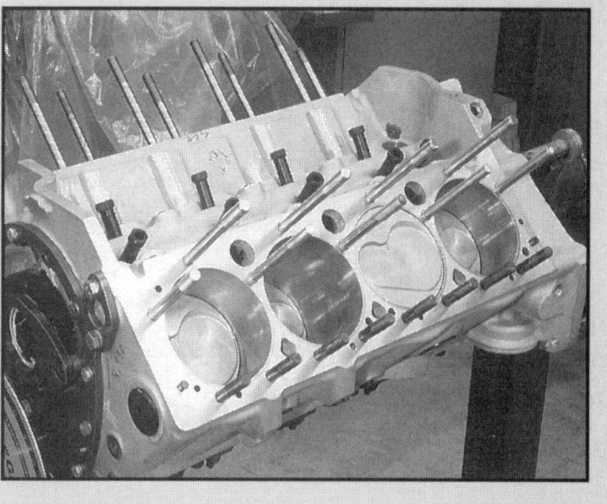

If you purchase a short block assembly and install your own heads, you might want to consider the following. Unless your crate motor is a turbocharged or supercharged engine that will see extended periods of severe service in a race car or a boat, there is little reason to use cylinder-head studs. Studs are needed to ensure even gasket crush and promote proper sealing in engines with very high cylinder pressures. The BMEP (Brake Mean Effective Pressure) in the average street engine rarely approaches the point where gasket integrity is threatened. Blown head gaskets are normally the result of improperly torqued bolts, detonation, or improperly faced surfaces on the head or block.

For street and bracket racing applications it is usually safe to use stock factory head bolts. Factory fasteners have the proper elasticity to provide effective clamping at the recommended torque specs. Aftermarket bolts—like those available from ARP, Manley, Milodon, and others—perform very well.

If you angle-cut the heads to gain compression, be certain to have the machinist spot face all the head bolt seats to make them parallel to the deck surface. Remember also that the bosses along the outside of the head for the short bolts can get very thin after angle milling. It is possible to break or crack the head in this area if care is not exercised when torquing the bolts.

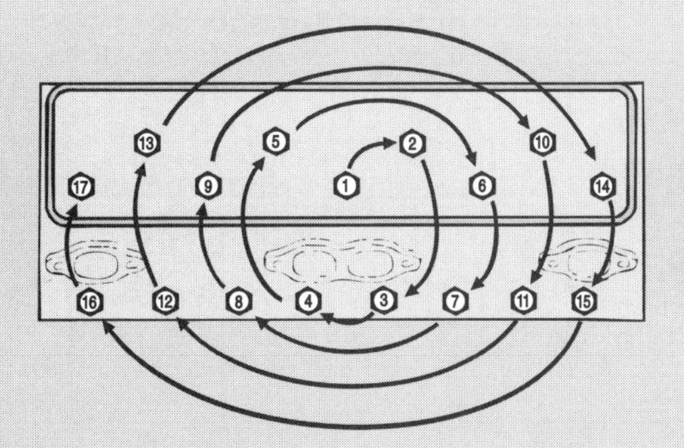

Follow the helical tightening sequence illustrated here for cylinder head bolts. First, torque all bolts to 25 lb-ft. Then repeat the process in 20 lb-ft. steps until you reach the full torque recommended for your fasteners (typically 65 to 75 lb-ft. for factory bolts).

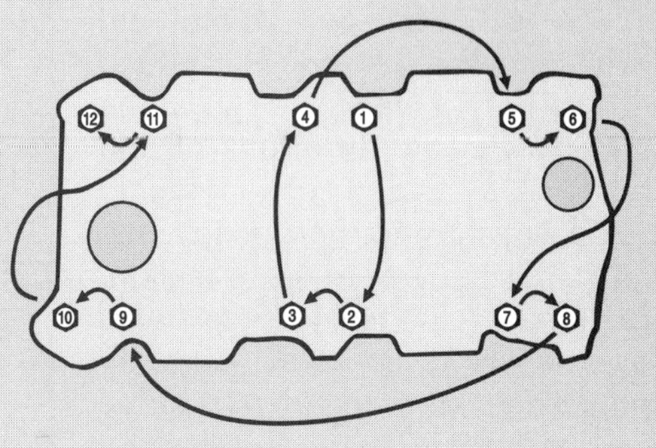

Here is a typical torquing sequence for the intake manifold. Tighten the bolts in 5 lb-ft. steps to a final torque of 30 to 35 lb-ft. As the gaskets compress, the bolts will loosen slightly. Retorque the fasteners until they remain tight.

ple-spark or long-duration ignitions almost universally help smooth out a rough idle and minimize plug fouling that can hurt engine performance during the first critical seconds after leaving the starting line. For the street, these high-tech ignitions can make stubborn starting a thing of the past, and improvements in gas mileage are not unusual.

SETTING THE TIMING

One of the most important requirements for getting maximum performance from your crate engine is stable accurate ignition timing. Even with a high powered ignition, the timing must still be accurately set and locked.

If you don't have a top-quality timing light, it's worth investing in one. Watch out for cheap units that have low light output; these strobes make it impossible to read the timing mark unless you're in complete darkness (and that's a great way to get your fingers chopped up in the fan blades). The best timing lights use an inductive pickup that quickly clamps on the outside of the ignition wire. Top quality lights will provide a stable indication of timing from idle to over 8000 rpm.

Checking initial timing with a strobe light is a relatively straightforward process. Connect the timing light to the number one spark plug terminal or wire according to the instructions provided with the light. Disconnect the manifold vacuum hose leading to the vacuum advance canister. Leave the hose connected to the manifold and plug the open end to prevent a vacuum leak. Start the engine and make certain the idle is below the point at which the centrifugal advance starts to activate. On most engines centrifugal advance may begin as low as 800 to 1000rpm. Point the timing light at the timing plate on the front of the engine. The flashing light will illuminate a timing mark on the spinning crankshaft damper and the stationary marks on the timing plate. If you have trouble seeing the marks, try enhancing them with a narrow stripe of white paint. The relative position

CRATE MOTOR Performance Tips & Tricks

IGNITION COIL BASICS

It all happens in the ignition coil. Composed of two separate wire windings (the primary and secondary) over a common iron core, the coil is a transformer that uses a magnetic field generated by the primary winding to produce a higher voltage in the secondary winding. The voltage increase is dictated by the ratio of turns in the primary versus secondary windings. The coil is located in the distributor cap on GM HEI distributors.

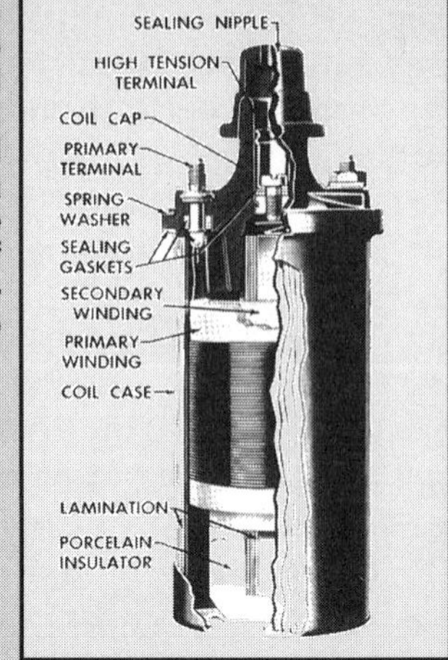

30,000 150

Many performance coils have a winding ratio of 30,000 turns/150 turns, or about 200 to 1. But this turns ratio would only increase battery voltage (12 volts) to about 2400 volts. Since ignition coils generate up to 60,000 volts, there must be other factors at work...read on.

The hidden factor is that voltage will be generated in proportion to how fast the magnetic field builds or collapses. If the magnetic field collapses quickly (in less than 1 millisecond) it will cause about 250 to 300 volts to be generated in the primary winding. This primary "spike," multiplied by the 200-to-1 turns ratio of the coil, produces up to 50,000 volts in the secondary that fires the sparkplugs.

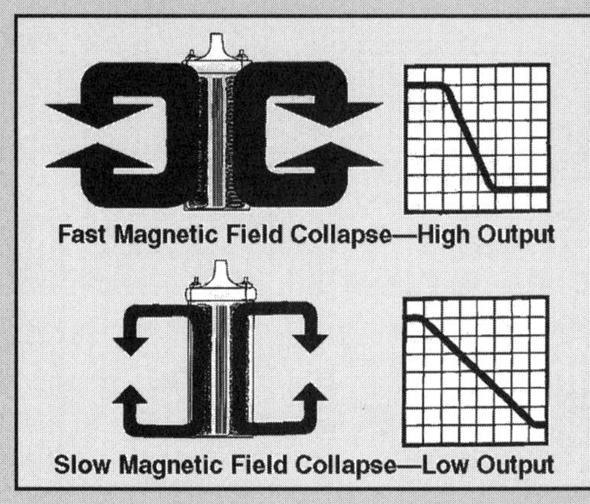
Fast Magnetic Field Collapse—High Output

Slow Magnetic Field Collapse—Low Output

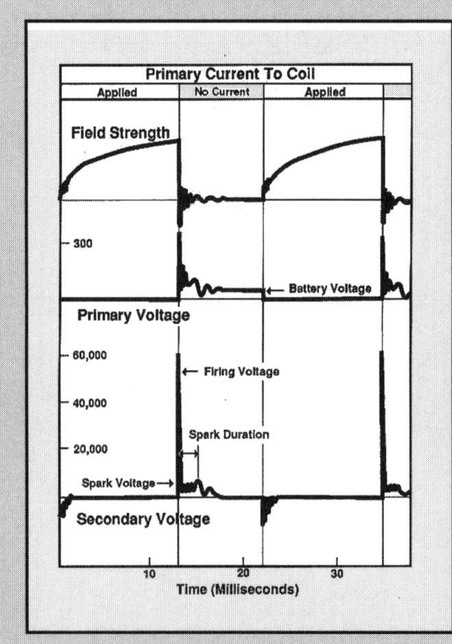

This is what happens to the magnetic field, primary voltage, and secondary voltage as power is applied and removed from the coil. When 12 volts is applied to the coil, the magnetic field builds. After several milliseconds the primary power is abruptly removed and the magnetic field collapses. This rapid change in the field generates nearly 300 volts in the primary and up to 60,000 volts in the secondary, firing the sparkplug.

of these marks will indicate the amount of initial ignition advance. In all but crank-trigger systems timing is adjusted by rotating the distributor housing (the pickup must be adjusted on crank trigger system to vary timing). Loosen the clamp securing the distributor and advance or retard (rotate) the distributor slightly until the timing light indicates the correct number of degrees (move the distributor counter-clockwise to advance timing on the small block). Once this is set, the distributor clamp should be firmly tightened.

This procedure is fine if you already know what the initial ignition timing should be for optimum performance. On a modified engine, the stock timing figure may no longer be applicable. Information presented in this chapter will help you find the best initial advance for your engine, but if you're just looking for a place to start, try setting initial at about 5 to 10 degrees advanced.

A timing light can also help troubleshoot the vacuum and centrifugal ignition advance mechanisms. By reattaching the vacuum hose to the advance canister while viewing the timing marks (make sure the vacuum hose is connected to manifold vacuum not a ported source on the carburetor), it is easy to confirm vacuum advance function. The timing mark on the vibration damper should move substantially ahead of the stationary TDC mark. In a similar fashion, the mechanical advance can be tested by simply increasing engine speed above idle (make sure the vacuum advance is disconnected for this test).

Finally, a quality timing light can help you determine how accurately the ignition system functions at high engine speed. Slowly increasing engine speed to near peak rpm while observing variations in the timing mark can reveal spark scatter, high-speed retard, or other mechanical or electronic abnormalities (and for safety reasons, never stand in line with the fan or fan belts—or remove the belts—when checking high-speed ignition timing). A performance ignition system should generate a rock-solid timing mark at all

CRATE MOTOR Performance Tips & Tricks

IGNITION CURVE REQUIREMENTS

If you are unsure about the best ignition advance curve requirement for your new crate motor, the following recommendations will help you establish a curve that will deliver maximum power.

Ignition Curve Requirements

Modification	Vacuum Advance	Mechanical Advance
Increase compression ratio	Retard overall curve (increase spring tension)	Less overall advance, especially at peak-torque (heavier springs and reduce advance weight movement)
Add a high-flow induction system	Remains unchanged	Less advance throughout RPM range (heavier springs and reduce advance weight movement)
Add headers	May need to either advance or retard, depending on charge temperature and exhaust contamination	If charge contamination is reduced, charge temperatures are usually lower. This often requires a slight increase in advance throughout the RPM range (Lighter springs and more advance weight movement)
Longer duration cam	Reduce spring tension in the diaphragm to bring the curve in faster	Initial advance needs to be faster and, to a lesser extent, total advance should occur at a lower RPM (both accomplished with lighter springs)
Supercharger	Needs special vacuum advance mechanism with pressure retard	Total advance should be reduced and the curve should come on slower (stiffer springs and less advance weight movement)
Turbocharger	Needs special vacuum advance mechanism with pressure retard	Reduce total advance with quicker initial part of the curve. Slow curve when turbo boost builds. (use a combination of light and stiff springs, plus reduce advance weight movement
Ported cast-iron heads	Remains unchanged	Slow high end of curve (add one heavy spring)
Aluminum heads	Reduced surface temperatures call for slightly more advance (increase advance amount and slightly reduce spring tension)	Often needs less total advance due to better cylinder filling, but faster initial advance due to lower port velocities and swirl (lighter springs and slightly reduce advance movement)
Switch to unleaded gas (same octane)*	Reduce overall advance (increase spring tension and amount of advance)	Reduce overall advance (increase spring tension and reduce advance weight movement)

*Note: Removing the lead in gasoline reduces the ignition delay time and requires reduced advance. Leaded fuels burn slower and require more advance.

CRATE MOTOR Performance Tips & Tricks

UNDERSTANDING IGNITION ADVANCE

Spark advance is the number of crankshaft degrees that the spark occurs before the basic (static or initial) spark timing. It has a great effect on how smoothly and efficiently your engine will run. In order to obtain the maximum pressure on the piston at the optimum position in the power stroke, ignition is often initiated well before the piston reaches the top of the compression stroke. However, as engine speed increases, there is less "real" time for the ignition process to be completed. Fortunately, flame speed increases nearly proportionally with engine speed—primarily due to the increase in combustion-space turbulence. This means that, for most of the rpm range, the crank angle through which combustion takes place can remain relatively constant. It is fortunate that flame speed is linked to engine speed. If this relationship did not exist, it would be impossible for spark-ignition engines to operate at high speeds because there would be insufficient time for efficient combustion to take place

The job of the various advance-control mechanisms, whether they are located in an electronic "black box" or controlled by various mechanical systems in a distributor, is to ensure that the spark is delivered to each cylinder at the correct time to gain maximum work (horsepower) from combustion. This should occur at all speeds from idle up to the rev limit. Timing advance is expressed as the number of crankshaft degrees before TDC (Top Dead Center) that the spark is delivered to the plug. A spark triggered after TDC is said to be retarded. Simply stated, the purpose of spark advance is to harness optimum power from the combustion process regardless of engine speed and load.

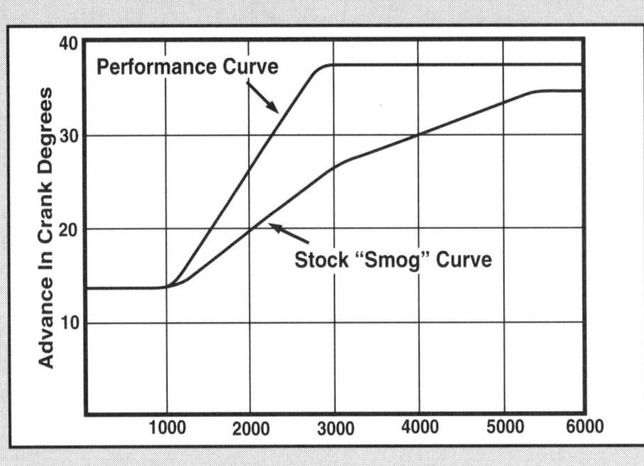

Set the initial timing about 12° to 16° advanced. Then adjust the advance curve to start about 1000 rpm. Have the curve add about 20 to 24 crank degrees by about 3000 rpm. The exact timing requirements will vary depending on the cylinder heads, transmission type, fuel octane and other factors.

INITIAL ADVANCE

Initial advance is the amount of fixed advance supplied to the engine during the start-up period. When the distributor is installed into the engine block and tightened in place, initial timing is fixed. Within the distributor the rotating breaker-point cam or reluctor is mounted on the centershaft. By loosening the hold-down bracket and adjusting the position of the distributor housing in the block, it is possible to change the relative position between the breaker plate and the centershaft, changing the initial timing at which the spark occurs relative to the crankshaft. Once set, the distributor clamp is tightened, locking the housing (and initial timing) in place.

MECHANICAL ADVANCE

While flame propagation speed increases with engine speed, this relationship is not directly proportional below about 3000rpm, especially in low-compression engines with low combustion chamber turbulence. From idle to 2500-3000rpm, the combustion rate increases much more slowly, and the centrifugal (mechanical) advance mechanism compensates by advancing ignition timing as engine speed comes up from idle. However as speed continues to climb, turbulence in the charge begins to speed flame propagation and reduce combustion time. This faster combustion rate offsets the need for further ignition advance. To match these variations in combustion time, most centrifugal mechanisms rapidly advance ignition timing in the lower rpm ranges, but as the engine speed builds timing is held constant or advances very slightly. If engine speed and "best" centrifugal advance are plotted on a graph, the curve usually resembles a steep slope up to about 2500rpm, followed by a flat plateau at higher speeds. To accomplish this, most smallblock distributors are equipped with bobweights and springs that move the breaker-point cam or reluctor relative to the distributor centershaft, advancing the ignition point as engine speed increases.

VACUUM ADVANCE

The other major factor affecting ignition timing is the variation in combustion time as the density of the air-fuel mixture changes. Since mixture density is much lower when the throttle is partially closed, cylinder pressure, turbulence, and flame speed are reduced. As a result, more time, in the form of additional ignition advance, is needed to burn the mixture. Since charge density and flame speed are directly related to manifold vacuum, a vacuum-sensitive mechanism is commonly used to increase or decrease advance, independently of the centrifugal mechanism.

Virtually all smallblock vacuum-advance mechanisms use a spring loaded diaphragm connected to the contact-breaker or magnetic-pickup plate with a linkage arm. At high vacuum levels, the diaphragm is retracted against the spring. This moves the breaker-plate in a direction opposite to that of the distributor centershaft, triggering the points or magnetic pickup sooner, advancing ignition timing. As the throttle is opened, vacuum decreases and the spring in the diaphragm assembly returns the breaker plate to its standard position, eliminating the additional timing advance that could otherwise cause ping or detonation at higher engine loads.

It is helpful to view the vacuum-advance system as a load-sensitive ignition control. Since the vacuum advance is progressively activated as power demands are reduced, e.g., during cruise, this system is the single most important ignition factor affecting fuel economy during normal street driving. If economy is at all important, the engine should be equipped with a fully functional vacuum advance.

TOTAL ADVANCE

Total advance at any rpm is the amount of timing lead supplied to the engine when all of the variable-advance systems have "corrected" spark timing. This includes initial advance put in by basic adjustment of the distributor housing, centrifugal advance supplied by the springs and weights in the distributor, and vacuum advance supplied by the vacuum canister.

CRATE MOTOR Performance Tips & Tricks

SETTING ROTOR TO CAP ALIGNMENT

When the coil is triggered and a high-voltage discharge is routed to the center of the distributor cap, this energy moves along the rotor to a single terminal on the cap and on to the appropriate spark plug...or does it? If the rotor is pointed directly at the correct terminal, everything works fine, but if the rotor is pointed halfway between two terminals, spark energy may find its way to the wrong plug, or to two plugs at once. These problems are due to incorrect rotor-to-cap alignment, sometimes called improper rotor phasing. Whatever you call it, it means lost horsepower and premature wear for the rotor, cap, and sometimes the ignition coil.

In the days of breaker-point ignition, one could rotate the distributor shaft and make a careful estimate when the points would just open, indicating the moment of spark discharge. Then holding the distributor shaft steady, it was possible to examine where the rotor was pointing in the distributor cap. This method was not the most accurate, although it did not require any special equipment. Now, most distributors use magnetic or other types of breakerless pickups that make it nearly impossible to visually determine at what distributor shaft position the coil will be triggered. So another method of measuring rotor-to-cap alignment must be used, preferably one that is accurate for all types of systems. This new method, in a way, is obvious: use a clear plastic cap and a timing light—or if a clear plastic cap is not available for your engine use a standard cap with large holes drilled in the top or side. A timing light directed at the cap will reveal the exact rotor position at the time of spark discharge.

The question is: where should the rotor be pointing? The most obvious answer—directly at the center of one of the distributor cap terminals—is usually correct, except when the distributor is equipped with a vacuum advance. In this case the rotor-to-cap alignment does not remain fixed, since as manifold vacuum increases, the advance canister moves the breaker plate to advance ignition timing, and moving the plate relative to the fixed distributor cap alters rotor-to-cap alignment. Distributors with vacuum advances must have the rotor alignment set off-center, or "late," in the direction of rotation when no vacuum is applied to the canister. This will ensure that when the vacuum advance is activated, rotor alignment will move past center to same position on the opposite, "early" side of the terminal.

Distributors that do not have vacuum advance canisters should have the rotor-to-cap alignment set "dead on," since the breaker plate does not move (it better not move!).

Correcting rotor-to-cap alignment can be difficult on some distributors and easy on others. On standard point or breakerless distributors, the breaker plate or pickup coil must be moved—rotated within the distributor housing—to alter rotor alignment. This sometimes involves drilling and tapping new holes for plate or pickup mounting. It is also possible to cut the metal blade in the rotor and solder it back pointing in a new position. However, the easiest way to set rotor alignment is to use an MSD adjustable rotor. By loosening two screws, turning the rotor, and retightening the screws, rotor misalignment can be corrected.

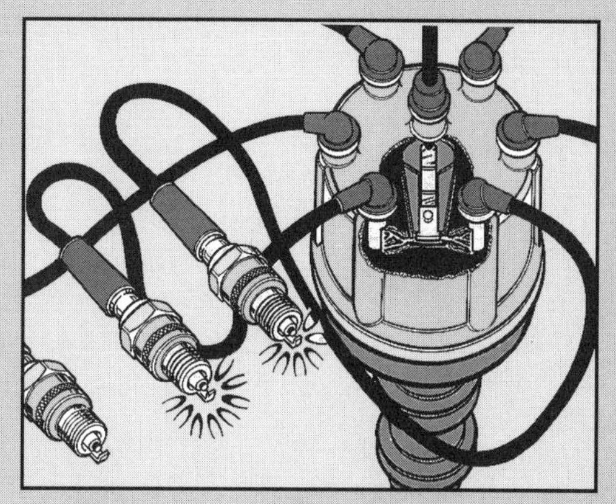

If the rotor is not pointed precisely at the proper terminal when the spark energy is released, the wrong plug, or two plugs at once, may be fired. This problem is called incorrect rotor-to-cap alignment, and it means lost horsepower.

The easiest way to check rotor-to-cap alignment is to drill a couple of holes in an old cap with a hole saw (or use a clear cap). Direct a timing light at the cap and you'll see the exact rotor position at the time of spark discharge.

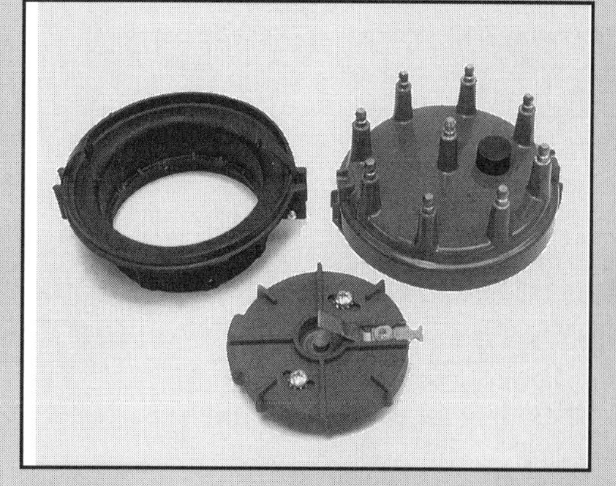

Correcting rotor-to-cap alignment once you've discovered it can be difficult with some distributors. You may have to drill new holes in the breaker plate and relocate the pickup coil or sensor. But if you install an MSD adjustable rotor (use Cap-A-Dapt if you don't have an HEI distributor), you can change rotor alignment in just minutes. Simply loosen the two lock screws (arrows), twist the rotor, and tighten the screws. Recheck alignment with a timing light.

engine speeds, there should be no visible signs of widening, spreading or jumping. If any of these problems are indicated, they can usually be traced to several mechanical and/or electronic sources, but the most common causes are a loose timing chain, worn distributor bushings, or a sticking mechanical advance. In addition, since the oil pump is driven off of the bottom of the distributor, spark scatter can often be traced to pressure "pulses" generated by the oil pump, especially when high oil pressure is used. Be careful to avoid this if possible.

The ignition system is a very important part of the crate motor horsepower picture. All of the work you may have done up to this point will be wasted if your ignition system doesn't generate reliable combustion. But don't be intimidated by the vast array of ignition equipment, and don't fall into the "over-complicated" pit. You may want to spend money on a quality breakerless ignition system to reduce maintenance or gain an added measure of high-rpm reliability, but don't believe that you must spend a lot of money. Don't cut corners, but also don't spend three times the money for equipment that doesn't add a single horsepower. You will achieve the greatest success through a careful blending of stock and specialty parts. Sort it out first, then reach for your wallet.

ADJUSTING VALVES

Most factory applications recommend that hydraulic lifters be adjusted one turn down from zero lash; the point where the lifter stops clicking while you're adjusting it. For maximum effect, cam manufacturers usually recommend an adjustment of only one-quarter turn. The procedure with the engine running is to loosen each adjusting nut until the lifter starts to click. Then tighten it until the clicking just stops. Once the clicking stops, slowly tighten the nut one-quarter turn. If you turn it down too fast, the lifter will not compensate and the engine will run rough or even stall, so work slowly.

If you're adjusting the valves in a new engine, the procedure is equally simple. All you have to do is place two fingers on the rocker arms just over the pushrods. Rotate the engine slowly until you feel them settle as far as possible. The rocker arms should be loose at this time. Continue turning the engine until the pushrods begin to rise. This means the lifters are off the base circle and they are beginning to move up the clearance ramp.

Work with one valve at a time and use the pushrod movement to determine when the lifter is on the base circle of the cam. Make the adjustment by turning down the adjusting nut slowly. While you are tightening the nut, grasp the pushrod between your fingers and slowly rotate it back and forth. When you have removed all the slack in the system, you can feel resistance as you twirl the pushrod. This means you are starting to depress the plunger in the lifter. At this point you go ahead and add your one-quarter turn adjustment and move on to the next valve. By presetting all the valves in this manner you will be able to start the engine and run in the cam without all the clatter normally associated with the initial fireup. The engine will start and run quietly; you can make a backup adjustment after all the parts have become familiar with each other.

This same basic method can be used with solid and roller lifters where you have to incorporate a certain amount of lash in the system. All you are really doing is making sure the lifter is on the base circle of the cam lobe when you make the adjustment. On an engine where the valves are already adjusted, the procedure is even quickest. Simply watch the rocker arms for each cylinder to determine the position of the lifters. Bump the engine around with the starter and when the exhaust valve just starts to open (rocker arm begins to depress valve spring), set the intake valve. Continue bumping the engine over until the intake valve is fully open. (Valve spring is fully depressed.) Watch it closely and when the valve starts to close (spring begins to

A degreed harmonic balancer can be helpful for setting and checking initial valve adjustment as well as ignition timing. Some crate motors come with degreed Fluidamprs as shown here.

rise), set the exhaust valve. This procedure ensures that the lifter is on the base circle of the cam lobe while the adjustment is being made.

Next to the cylinder heads, the camshaft and valve train represent one of the most important controls you have over your engine's output. There are remarkable power gains to be had from the right cam profile if it is well matched to the rest of your engine's components. Keep in mind that short duration, high-lift profiles are generally more conducive to good torque and smooth operation. By combining the correct cam timing with well integrated intake and exhaust systems, it is possible to achieve truly startling increase in economy and driveability. Select your camshaft on the basis of your intended driving style and the type of performance you require. And remember, nothing you can do will foul up your engine's performance potential as much as a horribly mismatched cam.

ENGINE LUBRICANTS

To break your engine in properly, you should use a non-detergent oil. Do not use a synthetic oil until after the engine has been fired and broken in. If you wish to switch to a synthetic oil at this time, be sure to drain the oil thoroughly and change the filter. If you are not switching to synthetic oil, simply change the filter and fill the pan with the correct amount of your favorite brand high performance motor oil.

One important lubrication issue is the initial startup of your new crate engine. This is a critical period where all the unfamiliar engine pieces need some time to to get to know each other. Maximum lubrication is essential. Prior to starting the engine it is a good idea to remove the valve covers and squirt some clean oil on the rocker arm fulcrums and on the valve stem/rocker arm contact point to ensure full lubrication.

Always pre-oil a new engine prior to starting it. Install the oil filter and insert a commercially available pre-oiling tool into the distributor drive hole to engage the oil pump drive (this one is from B&B Performance). Make up a simple pressure gauge that can be screwed into the rear oil gallery takeoff tap. Install a 1/8-inch plug in the front oil gallery takeoff. Attach an electric drill to the pre-oiling shaft and spin the pump to pressurize the system. Do this until oil flows from every rocker arm (it may take a couple of minutes).

Depending on your application, most engines can be pre-lubed with some type of commercially available pre-lube tool. This is typically a long drill motor attachment with the proper fitting to match up to the oil pump. This unit is lowered through the distributor hole and spun with a drill motor to turn the oil pump and pump oil through the engine. When you do this be sure to use a breaker bar to turn the crankshaft several times so that all the critical oiling passages on the rods and mains are exposed to pre-lube oil.

ENGINE STARTUP AND BREAK-IN

When you are finally ready to start you crate motor take a few brief moments to review everything you have done so far. You want the engine to start with minimal cranking so that you can take it up to about 2000 rpm right away to break-in the cam.

Start by reviewing all of your wiring connections. Is the starter circuit hooked up properly? Check for full battery voltage and make certain that the charging system is wired correctly and that the drive belts are properly routed and adjusted with the correct tension. Check the cooling system. Make certain the hose connections are tightened and that the engine are radiator are properly filled with coolant. Check your temperature gauge connections. Make sure it will monitor engine temperature during warmup and break-in so that you won't overheat the engine.

Check the oil level and recheck to see that the oil filter is correctly installed and tightened. check your oil pressure gauge connection. Make sure it is properly tightened and that it will read oil pressure or lack of oil pressure as soon as the engine fires.

Check all of your electrical connections for proper routing and make certain they are tight. Recheck the distributor to make sure it is properly oriented. It is very easy to install the distributor 180 degrees out of phase. To double check this, remove the number one spark plug and rotate the engine while holding your finger over the spark plug hole. When you feel compression in the cylinder it means you are approaching TDC. Observe the harmonic balance and timing mark and continue rotating the engine until the pointer indicates about 10° advance. This is sufficient to start the engine and break-in the cam. Install the distributor, allowing the drive gear to lead into the cam gear so that the rotor points to the terminal for the number one spark plug wire.

Snug the distributor, but leave it loose enough that you can rotate it with a little force. Check routing of all the spark plug wires for the proper firing order. Have the plugs been properly gapped? Hook up your timing light and arrange it so you can check the timing while the engine is warming up.

Check your fuel system connections. If you have an electric pump, turn it on and observe all connections for leaks. If you have a mechanical fuel pump, make sure the carburetor is primed. With the engine cold and ignition off you can fill the fuel bowls with startup fuel by trickling through the vent tubes with a small hose and a funnel. You only need enough fuel in the bowls to run the engine for a few seconds until the mechanical pump begins supplying fuel. If you have an electric pump make sure it is supplying the correct fuel pressure for your EFI system if your new crate motor is running electronic fuel injection.

Before starting the engine, have a friend operate the throttle to make sure it operates freely. If there is no interference, set the parking brake and place the transmission in neutral. Turn on the ignition and crank the engine. It should start after only a few cranks. If it doesn't start, turn off the ignition and review your preparations again.

When the engine starts, take it to 1800-2000rpm and hold it there. As it starts to warm up have a friend keep an eye out for fuel, oil and water leaks, If everything seems tight after a few minutes, set the idle stop to hold the engine at 2000 rpm. Let it run here for at least twenty minutes to break-in the cam. Keep an eye on it to make certain it doesn't begin to overheat.

Once the cam is run in, you can bring it back to idle and set the timing. Then drive the car to break it in. The best procedure seems to be normal driving with moderately strong acceleration, some idle time and some freeway time. Don't lean on the engine right away, but don't baby it either. After you get a few hundred miles on it you can feel safe in running it harder.

Most high performance crate motor applications are actually replacement engines being installed in compatible chassis where packaging is not a problem. Many others are actually engine swaps into street machines, street rods, off-road vehicles, tow vehicles and selected marine applications. In these cases engine fit and packaging can cause considerable grief if you fail to plan ahead. If this is your approach, you would be well advised to check out the available space carefully to make certain the engine will fit properly. The accompanying chart and dimensional diagrams show the physical dimensions of most popular crate motors. Compare these figures to the actual available dimensions in your engine compartment to ensure compatibility with your chassis. The old adage that, "you can put anything into anything," is generally true if you are willing to cut the fender wells, alter the frame, notch the firewall and leave off the hood. If that sounds a bit much for you, check your dimensions and select a more compatible engine for your high performance project.

Never forget the ultimate purpose of your crate motor application. If you a performing an engine swap using a crate motor, make sure the engine you select is compatible with your driving needs. For most applications, smoothness, drivability and reliability are far more important than power.

ENGINE DIMENSIONS CHART

Engine	CID	A	B	C	D	E	F	G	Oil Sump	Starter
Chevy V8	350/383/406	26.5	27	19.5	26	20.5	25	27	Rear	Left
Chevy V8	396/427/454/502	30.5	30.5	22	27	23.5	29.5	33	Rear	Left
Ford V8	302/351/400	27	29	20	22	22	25	27	Front	Right
Ford V8	427/460	30	32	23	27	28	30	32	Front	Right
Mopar V8	360	29.5	29.5	20.5	25	23.5	28	31	Front	Left
Mopar V8	383/400/440	29	30	23.5	29.5	24	28	30.5	Front	Left
Mopar Hemi	426	32	31	28.5	29	24	28	31	Center	Left
Olds V8	350/400	28.2	28.2	21.5	26	20.2	25	27.5	Rear	Right
Olds V8	455	29	31	22.5	26.5	24	27	31	Rear	Left
Pontiac V8	350/400	28.2	29	22	27	20	26	31	Rear	Left
Pontiac V8	455	29.5	32	23	27	27	28.5	33	Rear	Right

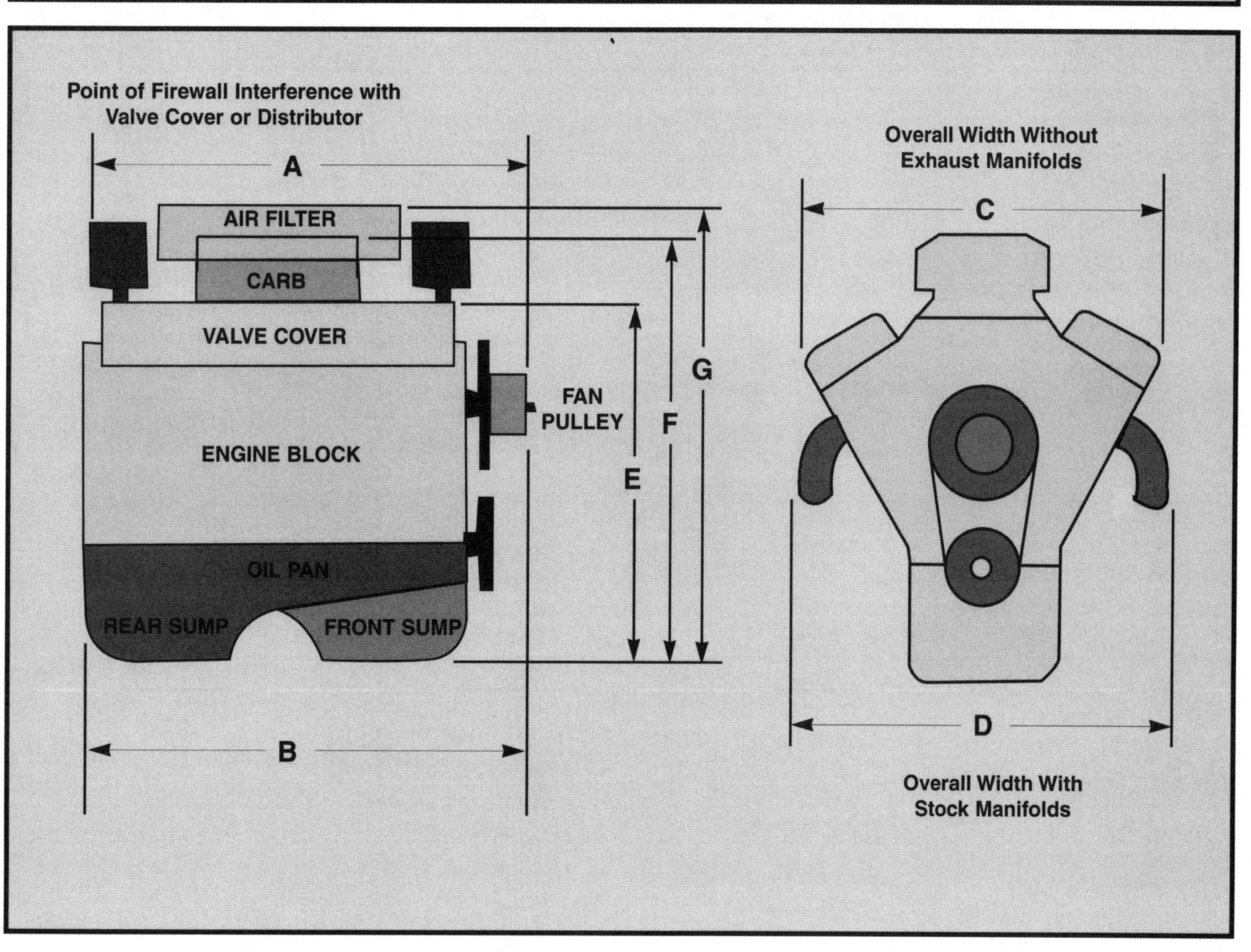

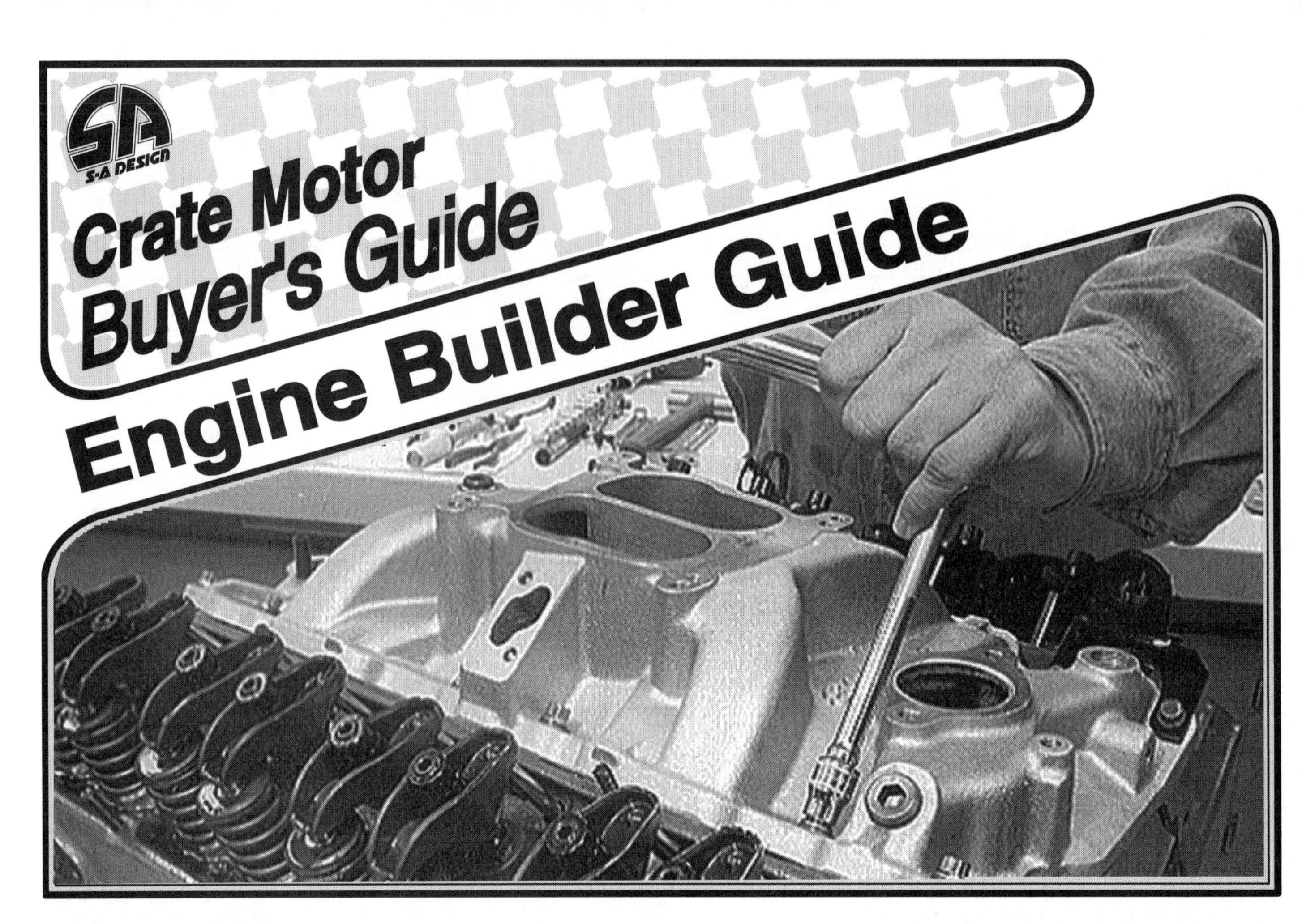

The following guide to engine builders is thorough and complete up to the date of publishing. Keep in mind however, that engine builders do move, change phone numbers and sometimes go out of business. The major crate motor engine suppliers shown in this book are all long term, well recognized engine builders supplying engines to the mail-order crate motor market. With rare exception, you can pretty much count on getting exactly what you ask for.

The author has little or no personal knowledge of the other engine builders listed in this section, hence he cannot vouch for the quality of their products or their business practices. It is assumed that the majority of them are capable of building a strong reliable performance engine to suit your requirements. You are urged to check out any engine builder prior to handing over cold cash. Reputable engine builders expect you to ask the tough questions.

ALABAMA

ALL PRO PERFORMANCE ENGINES
(205) 536-6959
2316 South Parkway
Huntsville, AL 35801

AUTOSPORT
(205) 533-2360
1018 Conception St
Huntsville, AL 35801

FAST AUTO PARTS CO INC.
(205) 263-2581
557 Bell St
Montgomery. AL 36103

HAMNER RACING ENGINES
(205) 424-4982
521 N 19th St
Bessemer, AL 35020

HUNTSVILLE ENGINE & PERFORMANCE CENTER
(205) 534-3181
Fax (205) 536 8905
315 E. Clinton Ave.
Huntsville, AL 35801

JORDAN PERFORMANCE CENTER
(205) 299-6636
Rt. 2, Box 225
Newton, AL 36352

MMADD RACING PRODUCTS
(800) 671 6995
2551 Balco Rd
Birmingham, AL 35210

ARKANSAS

AUTO SPORTS CENTER
(501)375-7223
Fax (501) 374-5385
2226 Cantrell Rd.
Little Rock, AR 72202

BAKER'S SPEED SHOP
(501)967-3821
Fax (501)967-3117
3203 Speed Shop Rd.
Russellville, AR 72801

BUNNS ENGINE REBUILDING
(501) 865-3079
Fax (501) 865-2737
Rt 1 Box 593
Donaldson, AR 71941

HALE PERFORMANCE WAREHOUSE
(501) 474-5252
Fax (501)474-3216
2209 Fayetteville Rd.
Van Buren, AR 72956

HOOD MACHINE COMPANY
(501) 425-3009
Rt. 8, Box 218H
Mountain Home, AR 72653

JACK MULLINS PERFORMANCE
(501) 572-7040
PO Box 2248
West Helena, AR 72390

KUNTZ & COMPANY
(501)246-2595
Fax (501)246 2696
540 Unity Rd.
Arkadelphia, AR 71923

ARIZONA

ARIZONA CUSTOM ENGINE SERVICE
(602) 842-1234
Fax (602) 842-2285
5314 W. Luke
Phoenix, AZ 85017

BRE PERFORMANCE CENTER
(602) 997-0800
Fax (602)582-1467
21616 N. Central Ave. #1
Phoenix, AZ 85024

BENNETT SPECIALTIES
(602) 269-6744
Fax (602)269-6204
3433 W Clarendon
Phoenix, AZ 85017

DJ'S DISTRIBUTING
(602) 497-0891
Fax (602)497-8068
2300 W. San Angelo, Suite 1093
Gilbert, AZ 85234

FAST EDDY'S ENGINES
(602) 233-3166
3031 N. 3151 Ave.
Phoenix, AZ 85017

GEMINI RACING SYSTEMS
(602) 940-9010
Fax (602) 940-9011
7301 W Boston St.
Chandler, AZ 85226

GORDIE'S SPEED CENTER
(602) 782-4744
Fax (602) 782-0794
1878 Arizona Ave.
Yuma, AZ 85364

HACKLEYS MACHINE SHOP
(303) 343-7692
425 E. 6th Ave.
Cottonwood, AZ 86336

KIP MARTIN RACING ENGINES
(602) 276-4763
Fax (602)276-4806
3533 E Corona Ave.
Phoenix, AZ 85040

KLEIN ENGINE DEVELOPMENT
(800) 845-2779
Fax (602) 967 7115
1207 N. Miller Rd
Tempe, AZ 85281

CALIFORNIA

FODGE & PECK MOTORSPORTS
(916) 635-6133
Fax (916) 635-7349
3263 Monier Cir, Unit H
Rancho Cordova, CA 95742

FOSSIL MOTORSPORTS INC.
(818) 709-0694
Fax (818) 718-8221
20715 Dearborn St.
Chatsworth, CA 91311

GAZAN RACING PRODUCTS
(909) 592-3235
908 N Cataract Ave
San Dimas, CA 91773

GOODSELL MACHINE
(619) 440-1070
1282 Pioneer Way
E1 Cajon, CA 92020

GREG SOLOW'S ENGINE ROOM
(408) 429-1800
Fax (408) 429-1801
125 Front St.
Santa Cruz, CA 95060

GREGG DAVIS COMPONENTS
(714) 692-7757
22345 La Palma Ave. #111
Yorba Linda, CA 92687

GRIGGS RACING PRODUCTS
(800) 655-0336
Fax (805)545-8260
285 Prado Rd., ~A
San Luis Obispo, CA 93401

HAIL'S AUTO MACHINE SHOP
(714) 871-2054
648 W. Williamson Ave.
Fullerton, CA 92632

HASSELGREN RACING ENGINES
(510) 524-2485
1221 4th St.
Berkeley, CA 94710

HIGH TECH PERFORMANCE
(818) 908-0943
15144 Raymer St
Van Nuys, CA 91405

HUFFAKER RACING, INC,
(707) 935-0533
28013 Arnold Dr.
Sonoma, CA 95476

I H P RACING ENGINES
(714) 990-6946
729 W Lambert Rd
Brea, CA 92621

JAE
(805)967-5767
375 Pine Ave
Goleta. CA 93117

JOHN PETERS RACING ENGINES
(707) 263-7151
Fax (707)263-1002
3817 Hill Rd
Lakeport, CA 95453

KEITH BLACK RACING ENGINES
(310) 869-1518
Fax (310) 869 2544
11120 Scott Ave.
South Gate, CA 90280

KOOPAL RACING ENGINES
(714) 597 2926
16050 Mountain Ave
Chino, CA 91710

LAMBERT ENTERPRISES
(805) 589-5491
12507 Old Town Rd.
Bakersfield, CA 93312

LOMA LINDA AUTOMOTIVE
(714) 796-0611
PO Box 941
Loma Linda, CA 92354

LOUIE UNSER RACING ENGINES
(714)879-8440
1100 East Ash Ave.
Suite C
Fullerton, CA 92631

A.E. RACING
(818) 768-2000
Fax (818)768 2312
7915 Ajay Dr.,
Sun Valley, CA 91352

ADVANCED ENGINE DEVELOPEMENT
(909) 279-8195
Fax (909)279-2627
149 N Maple St. Unit C
Corona, CA 91720

ADVANCED ENGINE MANAGEMENT
(310) 327-9336
Fax (310) 327-8520
15606 S, Broadway Ctr.
Gardena, CA 90248

ADVANCED ENGINEERING WEST
(909) 930-9852
Fax (909) 930-9756
1518 E. Francis St., Unit B
Ontario, CA 91761

AIR FLOW RESEARCH
(818) 890-0616
10490 Ilex Ave
Pacoima, CA 91331

ARAO ENGINEERING
(818) 709-4781
Fax (818) 709-4788
21400 Lassen St Unit G
Chatsworth CA 91311

AUTOCON ENGINEERING
(619) 279-0329
4333 Viewridge Ave. #B
San Diego, CA 92123

BESSANT ENGINEERING PRODUCTS
(714) 282-9275
205 E. Bristol Ln.
Orange, CA 92665

BPE RACING HEADS
(714) 572-6072
702 Dunn Way
Placentia, CA 92670

THE BALANCE SHOP
(818) 705-9046
7007 Darby
Reseda, CA 91335

CANADA

ANSWER ENGINE WORKS
(519) 472-3850
885 Sarnia Rd
London Ontario, N68 4B7
Canada

THE ATCHISON MACHINE SERVICE
(519) 451-3890
Fax (519) 659-4893
132 Clark Rd
London, Ontario, NSW 5E1
Canada

B S N AUTOMOTIVE
(519) 472-5610
885 Sarnia Rd
London Ontario, N6A 4B7
Canada

COMPETITION AUTOMOTIVE
(905) 889-0486
Fax (905) 889-0441
45 Maple Ave
Richmond Hill, Ontario,
L4C 6P3 Canada

GUYON RACING
(403) 277-6020
Fax (403) 277-4561
3702 Edmonton Trailine NE
Calgary, Alberta, T2E 3P4
Canada

LEITNER & BUSH ENGINEERING & PERFORMANCE
(905) 731-7026
Fax (905) 731-7027
93 Green Ln. Rear, Unit B
Thornhill, Ontario, L3T 6K6
Canada

KELLYS ENGINES
(519) 436-0730
PO Box 419
Blenheim Ontario, NOPIAO
Canada

JACKSON AUTOSPORT
(204) 235-11234
753 Marion Street
Winnipeg Manitoba.
R2JOK6 Canada

COLORADO

F & M AUTO PARTS INC.
(303) 433-8904
1729 N Federal Blvd
Denver, CO 80204

GORDEN MOTORSPORTS
(719) 597-0108
Fax (719) 597-0111
1330 Ford St.
Colorado Springs, CO
80915

LANIER'S SPEED SHOP
(719)633-1378
106 Pueblo Ave
Colorado Springs, CO
80903

CONNECTICUT

AUTO MACHINE SHOP
(203) 528-4184
660 Nutmeg Rd.
South Windsor, CT 06074

BOB BRUNEAU ENGINE RESEARCH
(203) 582-8164
Fax (203) 585-1580
43 Preston St,
Bristol, CT 06010

C B FABRICATION
(203) 886-8437
342 Norwich Ave
Taitville, CT 06380

CALLAWAY
(203) 434-9002
3 High St.
Old Lyme, CT 06371

DANBURY COMPETITION ENGINE
(203) 748-7356
88 Taylor St.
Danbury, CT 06810

HANK'S AUTO MACHINE
(203) 669-0241
204 E. Main St
Clinton CT 06413

FLORIDA

AEROSPACE COMPO-NENTS
(813) 545-4943
10681 75th St North
Largo, FL 34647

BENCO EXPORT INC.
(407) 845-9894
4261 Westroads Dr.
W Palm Beach, FL 33407

BILMAR
(904) 761-9904
1843 Taylor Rd.
Daytona Beach, FL 32124

BODIES AUTOMOTIVE
(813) 955-7749
2515 12th St.
Sarrasota, FL 34237

BOYD AUTOMOTIVE MACHINE & PARTS
(407) 425-3270
Fax (407) 425-8588
2748 S, Orange Blossom Trail
Orlando, FL 32805

CONSULIER INDUSTRIES
(407) 842-2492
Fax (407) 845-3237
2391 Old Dixie Hwy
Riviera Beach, FL 33404

THE CORVETTE CLINIC
(813) 571-1190
Fax(813) 573-1805
12295 A Automobile Blvd.
Clearwater, FL 34620

EMERALD COAST MOTOR-SPORTS
(904) 654 4351
Fax (904)654-7986
490 Holliday Rd
Destin, FL 32541

EXACT PERFORMANCE
(305) 942-6649
506 S. Dixie Hwy. E.
Pompano Beach, FL 33060

GREGG KRAMER RACING ENGINES
(305) 923-5756
2080 G Tigertail Blvd
Dania, FL 33004

HEADS BY RICK
(305) 983-1308
2129 SW 58th Terrace
Hollywood. FL 33023

J.B. RACING INC.
(904) 343-8900
13500 Southridge Industrial Dr.
Tavares. FL 32778

JVR ENGINEERING INC
(904) 796-8241
2475 Broad Street
Brooksville, FL 34609

KEITH EICKERT POWER PROD INC
(904) 446-0660
11 industry Dr
Palm Coast, FL 32137

GEORGIA

A & A RESEARCH
(404) 781-9437
3545 Dayion Dr
Cummings, GA 30130

BARNETT PERFORMANCE
(800) 533-1320
Fax (404)420-1572
465 Memorial Dr. SE
Atlanta, GA 30312

CALVIN NABORS AUTOMOTIVE
(404) 972-5469
1205 Old Snellville Rd
Lawrenceville, GA 30244

CHASTAIN ENGINE CENTER
(404) 478-7780
1065 Post Industrial Way
Jonesboro, GA 30236

D & A MACHINE SHOP
(404) 479-5873
764 Scott Road
Canton, GA 30115

DON DIXON HI PERFORMANCE
(404) 448-3166
2801 Cole Court
Norcross, GA 30071

ENGINE RESEARCH CO.
(404) 924-7859
1111 Shallowford Rd.
Marietta, GA 30066

ENGINE SYSTEMS
(404) 491-0583
Fax (404) 491-7357
2256A Fourth St.
Tucker, GA 30084

ERNIE ELLIOTT, INC.
(706) 265-1346
PO Box 476, Hwy. 183
Dawsonville, GA 30534

GLOBAL MOTORSPORTS
(404) 513-7500
Fax (404)513-3778
447 Gwinnett Drive
Lawrenceville, GA 30245

LAMAR HUNT AUTO INC.
(404) 974-9924
6362 Old Alabama Rd,
Acworth, GA 30102

LYLES SPEED & AUTO CENTER
(706) 652-2545
9029 Maysville Rd.,
PO Box 40
Maysville, GA 30558

IDAHO

DOUG HAYES ENGINES
(208) 628-3641
PO Box 499
Riggin, ID 83549

EXPRESSWAY MACHINE SHOP
(208) 233-3883
3941 Gateway Dr.
Pocatello, ID 83204

ILLINOIS

ADRIAN SPEED & MACH
(708) 231-9130
Fax (708) 231-9130
1225 W Roosevelt Rd
West Chicago, IL 60185

ALL PRO AUTOMOTIVE
(312) 237-0562
4836 W Fullerton Ave
Chicago, IL 60639

AMERICAN SPEED
ENTERPRISES
(309) 764-3601
Fax (309)764-2786
3006 23rd Ave
Moline, IL 61265

ANDRESEN ENGINE DEVELOPMENT
(815) 477-7223
13 Burdent Dr
Crystalake, IL 60014

AUTOMOTIVE MACHINE INC
(309) 347-7136
1000-04 Derby St
Pekin, IL 61554

B & B AUTO
(618) 252-2720
1621A Oglesbyrd
Harrisburg, IL 62946

THE BLOCK SHOP
(815) 695-5859
6 E. Front St,
Newark, IL 60541

BERTIL'S RACING ENGINES, INC.
(708) 395-4244
Fax (708) 395-5811
PO Box 438
Antioch, IL 60002

C P ENGINES
(618) 684-5035
Rt 3127 N
Murphysboro, IL 62966

CAR SHOP
(309) 797-4188
421 12th St.
Moline, IL 61265

THE CHEVY SHOP
(312) 777-1818
3517 N Cicero Ave
Chicago IL 60641

CHICAGO PERFORMANCE
(708) 455-8636
Fax (708)455-9086
9316 Franklin Ave.
Franklin Park IL 60131

D S S COMPETITION ENGINES INC.
(708) 268-1630
960 Ridge
Lombard, IL 60148

DAVE'S BALANCING
(217) 342-3643
801 E Fayette Box 1284
Effingham, IL 62401

DEAN MOTORS INC.
(708) 746-1900
3242 Sheridan Rd.
Zion, IL 60099

DUCKWORTH RACING ENGINES
(618) 288-6500
22 Schiber Ct
Maryville, IL 62062

ELITE ENGINES
(708) 395-1992
922 Carney Ct. Box 292
Antioch, IL 60002

ENGINE REBUILDERS
(815) 933-1978
RR 1 PO Box 28
Bonfield, IL 60913

FIVE STAR ENGINE SERVICE
(217) 525-1555
404 N Dirksen Pkwy
Springfield, IL 62702

GUSTAF AUTOMOTIVE & MACHINE
(309) 762-2323
4000 4th Ave.
Moline, IL 61265

HIXSON ENGINES
(217) 784-8623
314 E. 7th St.
Gibson City, IL 60936

HUGHES ENGINES INC
(309) 745-9558
23334 Wiegand Ln
Washington, IL 61571

HYPERCISION AUTOMOTIVE INC
(708) 532-0303
Fax (708)532-3044
15130 S Harlem Ave
Offand Park, IL 60462

KLEECO MOTORSPORTS
(309) 694-5007
Fax (309)694 3104
2244 E. Washington
E. Peoria, IL 61611

M.P.G. ACCESSORIES
(708) 432-2887
Fax (708)432 2861
2254 Skokie Valley Rd.
Highland Park. IL 60035

INDIANA

B E S RACING ENGINE
(812) 637-5933
78 Harrison Brookville Rd
West Harrison, IN 47060

ADVANCED ENGINE
(219) 784-8267
12549 Hwy 6
Plymouth, IN 46563

C.J. RAYBURN RACE CARS
(317) 535-8232
RR 1 Box 134A
Whiteland, IN 46184

CHAMPION AUTOMOTIVE MACHINE
(800) 442-2491
5002 W State Rd 234
McCordsville, IN 46055

DIXON RACING SUPPLY
(812) 246-4478
601 S. Indiana Ave
Sellersburg, IN 47172

ELDONS AUTO PARTS
(812) 372-2529
2161 State St
Columbus, IN 47201

ELEGANT MOTORS, INC.
(317) 253-9898
Fax (317) 257 3561
PO Box 30188
Indianapolis, IN 46230

FARMER'S AUTOMOTIVE
(317) 894-8185
2464 Buckcreek Rd.
Greenfield, IN 46140

GAERTE ENGINES
(219) 223-3016
Fax (219)223 8780
601 Monroe St
Rochester, IN 46975

GUNDERMANS PERFORMANCE CENTER
(317) 787-5144
Fax (317) 787-5337
4130 Madison Ave.
Indianapolis, IN 46227

HP COMPETITION ENGINES
(812) 522-2273
100 Ewing St
Seymour, IN 47274

HACKNEY'S PERFORMANCE SHOP
(317) 948-4666
223 S Walnut St
Fairmount, IN 46928

HARMON'S PERFORMANCE CENTER
(219) 368-7221
Fax (219) 368-9396
HiGhway 27 North
Geneva, IN 46740

HOGUE ENGINES INC
(219) 893-4161
1173S. 1075E
Akron, IN 46910

IDEAL RACING ENGINES
(219) 987-6543
800 15th St SE
De Motte, IN 46310

INDY CYLINDER HEAD
(317) 862-3724
8621 Southeastern Ave.
Indianapolis, IN 46239

JASPER ENGINES & TRANSMISSION
(800) 827-7455
PO Box 650
Jasper, IN 47547 0650

JONES ENGINEERING
(812) 254-6456
Rl 2 Sunnyside Rd
Washington, IN 47501

KATTERJOHN & ASSOC. PERFORMANCE
(812) 897-4956
4244 Hwy 62 W
BoonvHle, IN 47601

KERN'S SPEED SHOP
(812) 275-4289
203 N N St
Bedford, IN 47421

KINETIC ENGINEERING
(317) 272-3736
4729 Charles Dr
Plainfield, IN 46168

LINGENFELTER PERFORMANCE PRODUCTS
(219) 724-2552
1557 Winchester Rd
Decatur, IN 46733

M & M AUTOMOTIVE
(317) 353 1617
3024 S 5 Points Rd
Indianapolis, IN 46239

K-MOTION
(317) 742-8494
Fax (317) 742-1363
2381 N 24th St.
Lafayette, IN 47904

IOWA

AUTO RONS
(319) 324-0324
1029 W 4th St.
Davenport, IA 52802

AUTOMOTIVE ENGINE & MACHINE INC.
(319) 291-6569
123 Clark St.
Waterloo, IA 50703

BJ'S PERFORMANCE CENTER
(319) 372-8595
2514 Avenue L
Fort Madison, IA 52627

BENSKIN MOTOR SERVICE
(515) 753-8533
514 E. Anson St.
Marshalltown, IA 50158

GROVE AUTOMOTIVE
(319) 583-0757
Fax (319) 583-3435
3230 Dodge, PO Box 1306
Dubuque, IA 52004

HAP'S RACING ENGINES
(515) 232-4277
Fax (515) 233-5058
108 SE 2nd St.
Ames, IA 50010

HORN AUTOMOTIVE
(319) 377-0190
Fax (319)377 5500
850 50th St.
Marion, IA 52302

KIRCHNER'S
(319) 835-5084
Fax (319)835-9031
1868 Hwy. 2
Donnellson, IA 52625

L.A. ENGINEERING
(319) 276-4719
Fax (319)276-4719
1418 Easton Ave.
Waverly, IA 50677

KANSAS

CHARLIE WILLIAMS ENGINES
(913) 262-6330
2701 W 47th St.
Shawnee Mission KS 66205

DON PRESTON RACING
(316) 263-4769
Fax (316)263-7358
331 Pattie
Wichita, KS 67211

KENTUCKY

AUBURNDALE AUTO PARTS
(502) 368-3898
7114 Southside Dr.
Louisville, KY 40214

BLUEGRASS PERFORMANCE CENTER
(606) 734-4500
970 S. College St.
Harrodsburg, KY 40330

CORNETT RACING ENGINES
(606) 678-5163
Fax (606)679-5920
South Hwy. 27,
PO Box 127
Somerset, KY 42501

ESTES AUTOMOTIVE
(606) 252-1277
632 E Seventh St
Lexington, KY 40505

FRALEY AUTOMOTIVE INC
(606) 498-3440
Fax (606)498-3443
15 S Queen St
Mt Sterling, KY 40353

HEDGER BROS SPEED SHOP
(606) 491-5131
410 E 15th St
Covington. KY 41014

HERRICK MACHINE & PERFORMANCE
(502) 366-9185
Fax (502)366-9188
443 Downes Terrace
Louisville, KY 40214

J S A ENGINES
(502) 926-2458
Fax (502) 688 0432
715 E 4th St
Owensboro, KY 42303

JOE'S AUTO PARTS & MACHINE
(502) 366-0388
8000 Old Third St Rd
Louieville. KY 40214

LOUISIANA

BEN HATCHER RACING
(504) 340-4397
PO Box 305
Harvey, LA 70059

DIAMOND RACING ENGINES, INC.
(504) 345-9347
2480 Highway 190E
Hammond, LA 70401

MAINE

B & G SPEED SHOP INC.
(207) 767-2856
380 Lincoln St.
South Portland, ME 04106

MARYLAND

AHM PERFORMANCE
(410) 866-4747
7528 Philadelphia Road
Baltimore, MD 21237

ADVANCED AUTOMOTIVE
(301) 843-5700
2439 Old Washington Rd,
Watdorf, MD 20601

CRESAP AUTO MACHINE
(301) 724-8794
631 Mechanic St
Cumberland, MD 21502

EWART'S COMPETITION INC.
(301) 855-4549
637 Keith Ln
Owings, MD 20736

MASSACHUSETTS

ADVANCED RACING HEADS
(508) 373-8016
18 S Central St.
Bradford, MA 01835

B & C AND RPM
(508) 643-1362
90 George Levin Dr.
North Attleboro, MA 02670

BAKERACING
(413) 596-9475
2421 Boston Rd
Wilbraham, MA 01095

BOUCHER'S RACING ENGINES
(508) 948-7343
239 Haverhill St., Rte, 133
Rowley, MA 01969

CRE PERFORMANCE
(508) 355-2864
Fax (508)355-6364
Rt. 122
Barre. MA 01005

JKM AUTOMOTIVE INC.
(508) 966-2531
250 Farm St
Bellingham, MA 02019

MICHIGAN

A & G LIMITED
(313) 666-4283
6869 Desmond Rd
Waterford, MI 48329

ARROW RACING ENGINES
(313) 852-5151
Fax (810)852-8450
3811 Industrial Dr.
Rochester Hills, MI 48309

AUTO PARTS COMPANY
(616) 781-8650
13450 W Michigan Ave
Marshall, MI 49068

AUTO RACING'S TOTAL SERVICE
(517) 782-6237
955 Floyd Ave,
Jackson, MI 49203

AUTOMOTIVE TECHNIQUES
(313) 624-0200
220 Shamrock Hill
Novi, MI 48377

BACHMAN ENGINEERING
(616) 469-3837
25 S Willard
New Buffalo, MI 49117

BATTEN PERFORMANCE
(313) 946-9850
28884 Highland Rd
Romulus, MI 48174

DIXON PERFORMANCE ENGINES
(313) 261-4060
31362 Industrial Rd.
Lavonia, MI 48150

DOWKER ENGINES
(517) 543-0249
502 IsLand Hwy
Charlotte, MI 48813

FAST EDDIES MACHINE & PARTS
(517) 482-7599
4608 Northeast St
N US 27
Lansing, MI 48906

GIANINO RACE ENGINES
(313) 280-0240
4812 Leafdale
Royal Oak, MI 48073

HRP MOTORSPORTS INC.
(616) 874-6338
8775 Belding Rd.
Rockford, MI 49341

HUDSON AUTO CENTER INC
(517) 448-8986
123 W Main
Hudson, MI 49247

IMPASTATO RACING ENGINES
(313) 791-1660
Fax (313) 791-1660
19709 15 Mile Rd.
Mt Clemens, MI 48043

KATECH INC
(810) 794-4120
Fax (810) 791 0802
24324 Sorrentino Ct.
Clinton Township, MI 48035

KOFFEL'S PLACE
(313) 363-5239
4300 Haggerty Rd.
Walled Lake, MI 48390

LANDON RACING ENGINES
(616) 945-9620
5765 Usborne Rd.
Freeport, MI 49325

LIVERNOIS ENGINEERING
(313) 278-1221
Fax (313)278 5992
25851 Trowbridge
Inkster, MI 48141

MINNESOTA

ARCHER RACING PARTS & ACCESSORIES
(218) 727-1814
210 East 1st St.
Duluth, MN 55802

BALERS ENGINE SERVICE
(507) 282-5586
2915 20th St.
S.E. Rochester, MN 55904

CHRISTENSEN AUTO REPAIR
(612) 864-3764
1518 E 13th St
Glencoe, MN 55336

DYNO TUNES
(612) 425-8190
10550 County Rd. 81,
Suite 208
Maple Grove, MN 55369

EASTWOOD AUTO WORKS
(507) 282-3234
1225 Marion Rd SE
Rochester, MN 55904

HOME RUN ENGINE
(612) 762-8817
Fax (612) 762-8817
5413 County Rd 82 NW
Alexandria, MN 56308

INDUCTION RESEARCH
(612) 445-7821
1762 County Rd 18
Shakopee, MN 55379

JAX SPECIALTY ENGINES
(612) 478-9837
3568 Pinto Dr.
Hamel, MN 55340

JOHNSON AUTOSPORT
(507) 663-1210
Fax (507)663-1829
605 110th St. E.
Northfield, MN 55057

KELLEY AUTOMOTIVE
(612) 222-7374
359 S. Robert St.
St. Paul, MN 55107

MISSISSIPPI

COLUMBUS PERFORMANCE PRODUCTS
(800) 348-8032
Fax (601) 328-4454
5030 Hwy. 182 E,
Columbus, MS 39702

MISSOURI

ANDERSON RACING INC.
(816) 765-4881
3911 Main St
Grandview. MO 64030

BUCHER & SONS
(314) 742-2850
4005 E. Ridge Dr.
Pacific, MO 63069

CASPER RACING
(314) 631-8140
9269 Ummelman
St. Louis, MO 63123

CHECKPOINT ENGINEERING
(314) 968-4100
Fax (314) 968-0330
9331 Manchester Rd.
St Louis, MO 63119

DEPENDABLE AUTOMOTIVE
SERVICES INC.
(314) 423-8990
Fax (314)423-4350
10438 Lackland Rd.
Overland, MO 63114

FARLEY ENGINES
(816) 431-3550
Fax (846)858-3548
14025 92 Hwy.
Platte City, MO 64079

FRAME TECH
(314) 783-5533
508 Villar St.
Fredricktown, MO 63645

HALL RACING ENGINES
(314) 683-2280
Fax (314)683-2017
Beasley Park Road
Charleston, MO 63834

HARRY M, AUTOMOTIVE
(314) 343-9126
2141 S. Hwy 141
Fenton, MO 63026

HERZOGS MACHINE SHOP
(314) 867-6700
9844 W Florissant Rd.
St. Louis, MO 63136

NEBRASKA

CHARLEY'S SPEED AND MACHINE
(402) 426-9681
Fax (402) 426-9682
Hwy. 30 South
Blair, NE 68008

DALY'S MACHINE
(402) 330-3300
Fax (402)333-2595
13725 "C" St.
Omaha, NE 68144

GARY SELLIN CUSTOM ENGINES
(402) 371-4627
4007 Norfolk Ave
Norfolk, NE 68701

NEVADA

A-ALLIED SERVICE CENTER
(702) 364-0566
Fax (702) 364-1321
4047 West Desert Inn Rd
Las Vegas, NV 89102

NEW JERSEY

AUTOWERKS
(908) 968-5591
513 Jefferson Ave
Dunnelien, NJ 08812

BRUCE'S SPEED SHOP
(201) 335-0001
Fax (201)335-0971
3703 Rt 46
Parsippany, NJ 07054

BURNETT RACING ENGINES
(908) 359-2150
303 Griggstown Road
Belle Mead, NJ 08502

NEW MEXICO

ED'S ENGINES
100 Mesa Vista
Santa Fe, NM 87501
505 988-7504

ENGINE PARTS INC.
(505) 242-5285
Fax (505) 243-2156
900-902 Second St. NW
Albuquerque, NM 87102

FRITZ MARINE NGINEERING
(505) 445-9540
Fax (505)445-2211
1310 S 2nd St.
Raton, NM 87740

NEW YORK

ACTION PERFORMANCE
(516) 244-7100
1619 Lakeland Ave.
Bohemia, NY 11716

AL'S AUTOMOTIVE
(716) 731-3731
4663 Bear Rd
Ransomville, NY 14131

AMORES CUSTOM CARS INC.
(716) 373-1556
171 W Main St
Allegany, NY 14706

CAMPBELL ENGINES
(716) 335-7067
4944 Hackman Rd
Dansville, NY 14437

BILL MITCHELL RACING PRODUCTS
(516) 737-0372
Fax (516) 737-0467
35 Trade Zone Dr.
Ronkonkoma, NY 11779

CLIFTON PARK ENGINE REBUILDER
(518) 383-3092
1669 Rt 9 Box 367
Clifton Park, NY 12065

CARL MCQUILLEN RACING ENGINES INC.
(716) 768-2322
8171 E Main Rd
Leroy, NY 14482

ENDERS RACING ENGINES
(315) 655-8333
3211 Cazenovia
Nelson Rd.
Cazenovia, NY

EASTERN AUTOMOTIVE SUPPLY
(315) 453-3709
4003 E Bourne Dr.,
Box 5096
Syracuse, NY 13220

FORMULA MOTORSPORTS
(718) 482-0515
Fax (718) 482-0515
32 16 37th Ave.
Astoria, NY 11101

GORDEN AUTOMOTIVE
(716) 674-2700
60 N. America Dr
Buffalo, NY 14224
PO Box 108
Depew, NY 14043

JPM ENGINE REBUIL.DING INC
(914) 562-5796
135 S Robinson Ave
Newburgh, NY 12550

JACK MERKEL INC.
(516) 234-2600
1720 Expressway Dr S
Hauppauge, NY 11788

KIRKUM'S AUTOMOTIVE MACHINE
(716) 728-2540
11461 Rt 21 S
Way/and, NY 14572

NORTH CAROLINA

AK RACING ENGINES
(704) 455-6222
Fax (704)455 6224
6011 Victory Ln
Harrisburg, NC 28075

AUTOMOTIVE SPECIALISTS INC,
(704) 786-0187
Fax (704) 786-0190
65 Roberta Road
Concord, NC 28027

BARNES & REESE RACING ENGINES
(704) 252-5455
211 Amboy Rd.
Asheville, NC 28806

BILL DAVIS RACING, INC.
(910) 476-1114
Fax (910)476-6048
11 N. Robbins St.
Thomasville, NC 27360

BURKETT RACING ENGINES
(919) 527-7860
Rt 2 Box 260 B
Dover, NC 28526

BYERLY AUTOMOTIVE
(704) 869-2386
PC Box 154, Rt. 109 South
Denton, NC 27239

DOUG HERBERT PERFORMANCE PARTS
(800) 336-7652
Fax (704)435-0096
2405 Lincolntown Hwy, 154 E
Cherryvllle, NC 28021

HERB MCCANDLESS PERFORMANCE PARTS
(910) 578-3682
Fax (910) 578 5577
PO Box 741
Graham, NC 27253

OHIO

ADVANCED TECH
(513) 424-0101
6830 Chrisman Ln
Middletown, OH 45042

ALL PARTS & SUPPLY
(216) 454-7015
2214 Columbus Rd NE
Canton, OH 44705

ALL PRO ALUMINUM CYLINDER
HEADS INC.
(614) 967-7761
Fax (614) 967-9404
5370 Johnstown-Alex Rd.
Johnstown, OH 43031

AMERICAN ENGINE & WELDING
(513) 228-3001
600 W Third St
Dayton, OH 45407

AUTOCRAFT ENGINES
(800) 356-6586
1100 Custer Rd
Toledo, OH 43612

AUTOMOTIVE MACHINE SHOP
(513) 426-5275
3832 Dayton Xenia Rd
Beaver Creek, OH 45432

B T AUTOMOTIVE
(513) 797-7131
139 Mt Holly Rd
Amelia, OH 45102

BARBOURS MACHINE SHOP
(614) 353-1200
915 11thSt
Portsmount, OH 45662

BEACH PERFORMANCE
(614) 855-2143
10960 Johnstown Rd
New Albany, OH 43054

BECHSTEIN MACHINE
& FABRICATION
(419) 668-9670
191 Akron Rd
Norwalk, OH 44857

BELLE FONTAINE AUTO SUPPLY
(513) 592-3929
215 W. Columbus Ave
Belle Fontalne, OH 43311

BERBERICH AUTOMOTIVE
(614) 864-8215
1628 Rey New Albany Rd
Blacklick, OH 43004

BRAUR MACHINE
(216) 799-0607
3674 Mahoning Ave
Youngstown, OH 44509

COMER & CULP ENGINES
(513) 498-9879
1604 Wapakoneta Ave
Sidney, OH 45365

COMPTUNE INC
(216) 948-3338
Fax (216) 948 3338
7096 Lafayette Rd
Medina, OH 44256

COUNTY LINE PERFORMANCE PRODUCTS
(216) 466-5000
2888 N County Line Rd
Geneva, OH 44041

D & D PARTS & ENGINE
(513) 761-1030
6634 Vine St
Cincinnati, OH 45216

DRE
(216) 337-9319
1267 W Pinelake Rd
Salem, OH 44460

DAYTON CRACKED BLOCK SERVICE
(513) 222-6769
823 W 3rd St
Dayton, OH 45407

DICK HUNTLEY AUTO SUPPLY
AND MACHINE
(513) 544-2341
11540 State Rte. 41 South
W. Union, OH 45693

DON'S CARB AND AUTO CENTER
(216) 628-3354
3575 A Gilchrist Rd
Mogadore, OH 44260

DON'S CRANKSHAFT
(513) 241-6893
122 E Liberty
Cincinnati, OH 45210

DON'S SPEED SHOP
(216) 758-3850
6581 Appleridge Dr
Boardman, OH 44512

DONE RIGHT ENGINE
(216) 582-1366
12955 York Delta ~E
N Royalton, OH 44133

DOUG'S PERFORMANCE & REPAIR
(419) 298-3266
223 N. Michigan
Edgerton, OH 43517

ORAIME RACING ENGINES
(216) 837-2254
1300 S. Erie St.
Massilion, OH 44646

ELLISON ENGINE SERVICE
(513) 242-2230
6406 Vine St
Elmwood, OH 45216

EURAUTO
(513) 271-8077
5904 Bramble Ave
Cincinnati, OH 45227

FAST RACING
(614) 836-2503
555-1 Corbett
Groveport, OH 43125

FLEMING'S ENGINE SER-VICE
(614) 892-3295
Fax (614)892-3295
18000 Arrington Road
Utica, OH 43080

FOWLER ENGINE INC.
(614) 258-2924
Fax (614)258-8279
3021 Switzer
Columbus, OH 43219

GABLE MACHINE & ENGINES
(216) 724-9318
321 W Waterloo Rd
Akron, OH 44319

GEORGE'S
(513) 233-0353
Fax (513)236-3501
716 Brantiy
Dayton, OH 45404

GRIFF'S ENGINE & MACHINE
(419) 625-0783
710 Erie St
Sandusky, OH 44870

H & W AUTOMOTIVE PARTS
(419) 542-7771
Fax (419)542-6767
165 E High St
Hicksvilie, OH 43526

HOT SHOT MOTORWORKS
(419) 294-1997
Fax (419)294-1997
555 S Warpole
Upper Sandusky, OH
43351

HUCKS MACHINE SHOP
(513) 271-3676
5805 Madison Rd
Cincinnati, OH 45227

HUTTEN RACING ENGINES
(216) 285-2175
12550 Gat Hwy Box 309
Chardon, OH 44024

IMPAC HI PERFORMANCE
(419) 726-7100
5515 Enterprise Blvd
Toledo, OH 43612

LAKOCY'S ENGINE PARTS & MACHINE SHOP
(216) 323-8110
130 1/2 Clevelarrd St
Elyria, OH 44035

LARRY'S ENGINE SERVICE
(513) 241-4450
1676 Central Ave
Cincinnati, OH 45214

LLOYD BROTHERS OFF ROAD CENTER
(419) 878-5693
307 South St,
Watervii/e, OH 43566

M C L MACHINING
(513) 927 5555
3559 St Rt 136
Hiilsboro, OH 45133

BUDGET ENGINE REBUILDERS
(216) 281-4040
Fax (216)281-0589
3766 Ridge Rd

GELLNER ENGINEERING
(216) 398-8500
2827 Brookpark Road
Parma, OH 44134

OKLAHOMA

BRAND RACING ENGINE
(405) 745-3332
7904 S Council
Oklahoma City, OK 73169

CARBONE RACING ENGINES
(918) 835-6596
3907 E Admiral Pl
Tulsa, OK 74113

LEN WILLIAMS AUTO MACHINE
(918) 836-7583
739 S. 83 E. Ave.
Tulsa, OK 74112

OREGON

ANDREST PERFORMANCE
(503) 878-2146
21882 Hwy. 62
Shady Cove, OR 97539

IVEY ENGINES INC.
(503) 255-1123
Fax (503) 255 6337
4722 NE 148th St.
Portland, OR 97230

PENNSYLVANIA

APPLE CHEVROLET
(717) 843-8017
Fax (717)843-5730
Rt 30 & Roosevelt Ave,
PO, Box 7326
York, PA 17405

AUTOMOTIVE MACHINE SPECIALTIES
(215) 326-9926
Fax (610)326 1314
337 W High St.
Portstown, PA 19464

CHUCK'S PERFORMANCE
(412) 295-3075
131 Locust Dr
Freeport, PA 16229

DAVE GEORGE RACING ENGINES
(215) 767-6683
5419 Rt 873
Schnecksviile, PA 18078

DOMHOFF'S AUTOMOTIVE
(412) 452-8241
205 Old Saltworks Rd
Harmony, PA 16037

DON OTT RACING ENGINES
(717) 259-9440
265 Peepytown Rd.
East Berlin, PA 17316

ETR MACHINE
(814) 345-4166
Rd, 1 Box 455
Morrisdale, PA 16858

ELMERS ENGINES
(412) 935-4830
3782 Wexford Run
Wexford, PA 15090

EMPSON AUTOMOTIVE
(814) 326-4547
211 Main
Knoxville, PA 16928

EXECUTIVE AUTO SPORT INC,
(2152) 559-9232
247 N 3rd St.
Easton, PA 18042

H & H MACHINE
(814) 563-4562
Rd 2 Box 221
Pittsfield, PA 16340

HG ASSOCIATES INC.
(610) 495-6674
PO Box 5052
Limerick, PA 19468

HORST REPAIR
(717) 375-4690
Fax (717) 375-2708
694 New Franklin Rd
Chambersburg, PA 17201

ISSI BROS PERFORMANCE
(412) 489 9006
150 State St
Charleroi, PA 15022

J & L PERFORMANCE
(717) 292-4471
1735 Alpine Rd.
Dover, PA 17315

JENKINS COMPETITION
(215) 644-9328
Fax (215) 644-3390
153 Pennsylvania Ave
Malvern, PA 19355

KREM SPEED EQUIPMENT INC,
(814) 724 4806
Owner. A/an Krem
Rd 5 Box 170
Meadville. PA 16335

KRINER'S ENGINE BALANCING
(717) 263-8051
512 Elm Ave
Chambersburg, PA 17201

RHODE ISLAND

KELLY RACING ENGINES
(401) 941-7787
75 Russe St,
Cranston, RI 02910

SOUTH CAROLINA

M & H ENGINEERING
(803) 794-4434
1704 Crapps Ave
West Columbia, SC 29169-5506

CHARLES POWELL ENTERPRISES
(803) 871-8331
330 Robbin St
Moncks Corner, SC 29461

GENE'S SPEED SHOP
(803) 224-6214
2513 Stanridge Rd.
Anderson, SC 29625

MPG AUTOMOTIVE
(803) 882-5319
Fax (803) 224 3719
1004 East North 1st St.
Seneca, SC 29678

SOUTH DAKOTA

ALL MOTORSPORTS
(605)348-0695
1840 Lombardy Dr.
Rapid City, SD 57701

CORBETT MIDDLEN PARTS
(605) 338-5034
Fax (605)332-2467
2520 S Minnesota Ave
Sioux Falls, SD 57105

TENNESSEE

ADVANCED RACING TECH
(615) 472-8528
2419 Thompson Ln
Cleveland, TN 37311

CUSTOM RACE ENGINES
(615) 573-1449
2010 W John Sevier Hwy
Knoxville, TN 37920

EAGLE RACING ENGINES
(615) 524-2154
1531 N 6th Ave.
Knoxville, TN 37917

ENGINE TECH RACING ENGINES
(615) 877-5700
Fax (615)875-4666
3656 Hixson Pike
Chattanooga, TN 37415

HUTTON MOTOR ENGINEERING
(615) 648-3333
Fax (615) 648-1119
1815 Madison St.
Clarksville, TN 37043

LINK AUTOMOTIVE SERVICE
(615) 256-3060
1229 Lebanon Rd
Nashville, TN 37210

TEXAS

AUTO CENTER INC.
(214) 634-3900
9001 Sovereign Row
Dallas, TX 75247

CLEM COMPETITION ENGINES
(214) 503-8044
10918 Alder Circle
Dallas, TX 75238

COMPETITION PARTS
(713) 697-3400
Fax (713)697-2236
7000 North Freeway, Suite 300
Houston TX 77076

DELEON AUTO ACCESSORIES
(210) 682-1652
Fax (210) 630-3350
2625 S. 23rd St.
McAllen, TX 78503

FILLIP AUTOMOTIVE
(915) 949-2333
Fax (915) 944-1098
9543 S US Hwy 67
San Angelo, TX 76904

JOE WHITE'S PERFORMANCE AUTO
(512) 442-8423
4313 Gillis
Austin, TX 78745

KENDRICK AUTOMOTIVE
(210) 824-8020
8703 Botts Ln
San Antonio, TX 78217

UTAH

BAKER MACHINE
(801) 753-3190
90 West 2500 North
Logan, UT 84321

FUEL AIR & SPARK TECHNOLOGY
(801) 776-0437
44 W 2400 N
Sunset, UT 84015

HUNTER MACHINE
(801) 964-6179
Fax (801)964 0475
4669 West 3500 South #6
West Valley City, UT 84120

JOHNSON MACHINE SHOP
(804) 298-2142
375 South 200 West
Bountfful, UT 84010

VERMONT

A.C. PERFORMANCE CENTER
(802) 655-3818
8 Roosevelt Hwy. Rt. 7
Colchester, VT 05446

ALSUP RACING ENGINES
(802) 457-3111
Owner, Bill Alsup
60 Pleasant St.
Woodstock, VT 05091

HEADS UP MOTORSPORTS, INC.
(802) 886-2636
Fax (802)886-2675
RR 4, Box 135A
Chester, VT 05143

VIRGINIA

AUTOTHORITY, INC.
(703) 323-7830
3763 Pickett Rd.
Fairfax, VA 22031

C & C PERFORMANCE
(703) 368-7878
8081 Centerville Rd.
Manassas, VA 22111

DINWIDDIE AUTO PARTS
(804) 469-3444
13408 Boydton Plank Rd
Dinwiddie, VA 23841

DON'S PRECISION HEADS
(703) 491-4323
1030 C Cannons Ct
Woodbridge, VA 22191

FLETCHER HARRISON RACING
(804) 861-9184
230 Franklin St,
Petersburg, VA 23803

H & E RACING ENGINES
(804) 798-0384
145 Dow Gil Rd.
Ashland. VA 23005

HI-PERFORMANCE HARDWARE
(703) 533-8403
Fax (703)533-9733
5912 N. Washington Blvd.
Arlington, VA 22205

HUDSON AUTO MACHINE SHOP
(703) 343-4062
227 Walnut Ave
Vinton, VA 24179

KEVIN BLANKS PERFORMANCE INC.
(804) 374-2188
Fax (804) 374-0441
PO Box 1811, Hwy. 58 West
Clarksvilie, VA 23927

LEE EDWARDS RACING
(703) 788-4210
Box 235, Rd, 616
Calverton, VA 22016

WASHINGTON

FLOW TECHNICS
(206) 697-1713
13310 Lakeshere Dr NW
Poulsbo, WA 98370

HONEST PERFORMANCE
(206) 838-7070
33210 Pacific Hwy. S.
Federal Way, WA 98003

JIMGREEN'S PERFORMANCE CENTER
(206) 774 3507
6824 212th SW
Lynwood, WA 98036

K.B. INDUSTRIES SPEED EQUIPMENT
WAREHOUSE
(509) 482-2926
E. 1614 Holyoke
Spokane, WA 99207

A C'S SPECIALTIES
(206) 256-5787
13917 NE Fourth Plain
Vancouver, WA 98682

WEST VIRGINIA

DAN COOK ENTERPRISES
(304) 387-3203
149 Fairview Rd.
Chester, WV 26034

WISCONSIN

AUTOMOTIVE MACHINE SERVICE
(414) 774-7005
9426 W Schlinger Ave
West Allis, WI 53214

B+B RACE CAR ENGINEER-ING
(414) 739-2657
Fax (414) 739-1550
2588 Coldspring Rd
Appleton, WI 54915

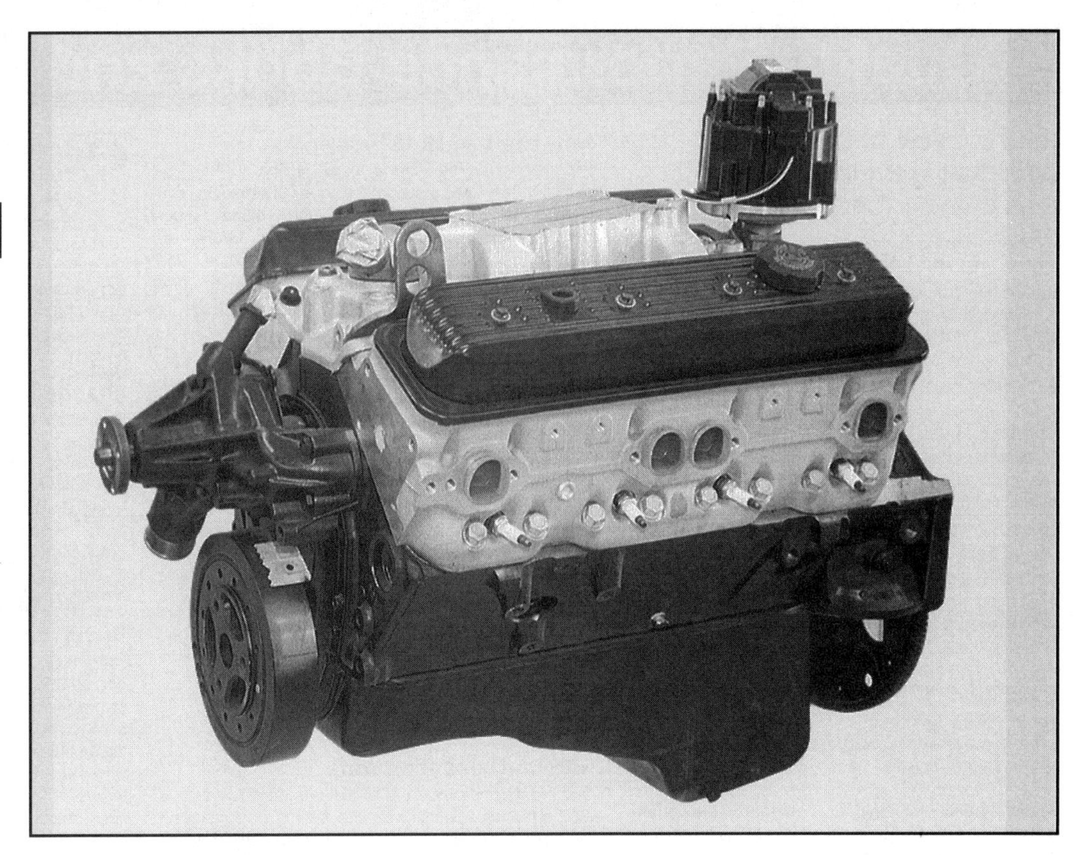

BARIL ENGINE REBUILDERS
(414) 336-4600
Fax (414)336-4700
996 Centennial St
Green Bay, WI 54304

BAY SPEED CENTER
(414) 336-8261
2587 Lawrence Dr.
De Pere, WI 54115

BRUNTON MOTOR PARTS
(608) 365-0104
Fax (608)365-7916
510 Broad St
Beloit, WI 53511

C & S PERFORMANCE
(414) 781-0469
12420 W. Derby Pl.
Butler, WI 53007

EAGLE PERFORMANCE
(608) 847-7869
Rt, 3, Powers Ave.
Mauston, WI 53948

G & H AUTOMOTIVE
(414) 685-6780
6191 Hwy 21
Ornro, WI 54963

KELLY-MOSS MOTOR-SPORTS INC.
(608) 274-5054
3017 Perry St.
Madison, WI 53713

KING MOTORSPORTS INC.
(414) 593-2800
Fax (414) 593-2627
105 E Main St
Sullivan. WI 53178

Quality Performance Books and Software

Here's A Few Of Our Best-Selling Books-Call For Your Free Catalog Today!

The Chevrolet Racing Engine

An inside look at how master engine builder Bill "Grumpy" Jenkins squeezes 700 hp from a carbureted smallblock. Many experts refer to his guide as "the bible of racing engines" Part No. 01

Smokey Yunick's Power Secrets

The "inside story" from Smokey Yunick, a living legend in racing circles. Smokey discusses performance from carburetors to oil pans. Written in the direct style that has become his trademark. A truly unique performance book. Part No. 06

Smallblock Chevy Performance

Vol. 1: 1955 to 1981

Revised edition explores *every* important aspect of the classic smallblock Chrevrolet. Chapters detail parts selection, lower-end, pistons, cylinder head prep, ignition, carbs and much more! Covers all 1955 to 1982 engines. Part No. 07

Smallblock Chevy Performance

Vol 2: 1982 & Later

Late model edition of our bestselling Chevy parts selection guide. Includes the latest factory and aftermarket components. Details lower-end, pistons, cylinder heads, induction, ignition, camshafts, valvetrain, exhaust, and much more! Part No. 14

Building the Smallblock Chevrolet/Step by Step

Follow the steps in this *Workbench®* book shows the techniques and tricks the pros use to assemble an engine from the oil pan to the carburetor. Includes hundreds of performance tips and much more! A "must-have" for any enthusiast. Part No. 35

Dodge/Plymouth Performance

Packed with clear illustrations, this easy-to-read guide show how to build power and torque from the renowned A-series smallblock to the monster B-series big-block; for street performance, racing, or economy. Part No. 03

Ford Performance

A complete performance reference for the small-block, big-block, and Cleveland engines. Filled with state-of-the-art tech tips form top racers and engine builders. Our readers say this is the best Ford engine book ever published. Part No. 05

V-6 Performance

An in-depth guide to building V-6 power. Experts explain performance tricks for Chevrolet, Buick, and Ford engines that can produce enough hair-raising power to satisfy most rabid road warrior. By expert writer Pat Ganahl. Part No.13

Engine Blueprinting

The author's acclaimed combination of savvy writing and wrenching skills put this book in a class by itself. Includes: using tools, block selection and prep, crank mods, rods, pistons, heads, cams, assembly tips, and much more! A superb book. Part No. 21

How to Build Horsepower-Vol. 1

Discover how expert David Vizard builds winning engines for the street or track. Includes: shortblocks, cylinder heads, cams, inductions, ignition systems, carburetion, headers, exhaust, and much more! Learn how the pros build horsepower in virtually any engine. One of our all-time best sellers! Part No. 24

Holley Carburetors-Selection & Tuning

An easy-to-read guide for selection, installing and tuning all popular Holley performance carburetors. Clearly explains basic carburetor function, tuning for street, racing, off-road, turbocharging, economy, and more! Hundreds of photos and drawings. Part No. 8

Rebuilding Holley Carburetors/Step-By-Step

Takes you step-by step through disassembly of all popular Holley modular carburetors. *Workbench®* format makes rebuilding a snap, details popular performance enhancements, troubleshooting, modifications, and much more! The ultimate Holley book! Part No. 27

Carter Carburetors

The only authoritative source for tuning and rebuilding Carter four-barrel performance carbs. Considered an outstanding reference by experts; loaded with details that are difficult to find for the Thermo-Quad, AVS, AFB and WCFB carburetors. Part No. 11

Nitrous-Oxide Injection

Find out why nitrous is the simplest and easiest way to add horsepower. Includes the basics plus in-depth knowledge available from no other source. Packed with dyno and product test, bolt-on tops; it's all here! Don't buy or use nitrous without it. Part No. 16

Street Supercharging

An award-winning guide to modern supercharging science, from basic design to hands-on installation tips. Explores virtually every type and brand of superchargers, method of drive, performance potential, cost, etc. Includes listing of sources and accessories. By expert Pat Genahl. Part No. 17

Performance With Economy

Find out how to build performance an d still get good gas mileage. Includes: choosing exhaust (including muffler selection) and intake systems, turbocharging, water and nitrous-oxide injection, high compression, octane boosters, and more. No gimmicks, just accurate, dyno-tested info. Part No. 09

Propane Fuel Conversions

An all-new book that shows how to improve horsepower and reliability by using *propane*, a unique high-octane fuel. Details installation, how to run both gasoline and propane, turbocharging, avoiding mistakes, troubleshooting, and a lot more! Includes list of manufacturers, sources. Part No. 12

Do-It-Yourself Custom Painting

Produce custom paint tricks in your own driveway! Includes: practical advice on paint selection, surface prep, repainting techniques, flame painting, murals, pin-striping, and more. Do it with Custom Painting! Part No. 10

NEW - Performance Modifying Chevy Trucks for Street Strip and Off-Road

Coverage includes all full-size and mid-size Chevy trucks from 1975 to the present and provides in-depth how-to information on all aspects of modifying these trucks to meet personal tastes and uses. Includes raising and lowering suspensions, engine upgrades, factory part specs and interchanges. Includes 4X4's, 2X4's, S-Series and C-Series. Part No. 30

NEW - Chevrolet Big Block Parts Interchange Manual

Beginning with the earliest big blocks through present models, this first ever work provides complete part interchange information, allowing the hot rodder to custom build his own high performance version of the famous Chevrolet "Rat" motor from off-the-self factory parts. Includes factory part numbers, casting, marks, production histories, suppliers, performance capabilities of various components, etc. Part No. 31

NEW - High Performance Crate Motor Buyer' s Guide

Now days, the automobile manufacturers themselves and aftermarket builders have gotten into the business of producing and selling race ready and high performance engines. This book explains these engines to the enthusiast, what components do and don't come with the engine, how the engines are built, power and torque outputs, which engines best fit which applications, costs, problems and shortcomings about which to be concerned, legal issues, etc. (Avail. April 1996) Part No. 32

NEW - How to Design and Install High Performance Car Stereo

First ever book in the marketplace which provides the enthusiast and car lovers alike with expert knowledge and up-to-date details regarding design and installation of stereo componenets. In addition to providing do-it-yourself information, this work will also leasd the reader to making and intelligent purchase of an entire system from a specialty shop. Part No. 45

Quality Software For The PC from Motion Software

DeskTop Dyno

This software for your PC is, simply, a leap beyond all other engine building/testing simulations. Use simple pull-down menus to "assemble" any one of millions of engine combinations - literally in seconds! Then "run your test engine through a powerful full-cycle simulation and watch the DeskTop Dyno draw incredibly accurate horsepower and torque curves. Fun , easy-to-use, accurate, *and it cost less than all the rest!* Test 1 to 12 cycle engines, from 17 to 1000+ cubic inches, with smallblock, big-block, or 4-valve cylinder heads and more! Requires IBM compatible PC computer 3 1/2-inch disk drive (exchangeable for 5-1/4-disks if needed), includes easy-to-read 20-page manual. Turn your computer into $50,000 dyno test cell for only $39.95! Fully guaranteed. Part No. 40

HOW TO ORDER: Please purchase S-A Design products from your local speed shop, book store, or auto parts outlet. However, if you can not obtain our books locally, you may order directly from CarTech by calling us at 1-800-551-4754 (within the US) or at 612-583-3471.

Join Our Book Users Group - It's FREE!

Join the CarTech **B**ook **U**sers **G**roup now! Just some of the benefits are; 1) special discounts on books, 2) information of the latest publications. Join *BUG* and stay in touch - and it's free!

TO SIGN UP AS A BUG MEMBER
CALL 800-551-4754 or FAX 612-583-2023